Two-Dimensional Nuclear Magnetic Resonance in Liquids

Two-Dimensional Nuclear Magnetic Resonance in Liquids

by

AD BAX
Colorado State University,
Department of Chemistry

Delft University Press

D. Reidel Publishing Company
Dordrecht, Holland / Boston, U.S.A. / London, England

Library of Congress Cataloging in Publication Data

CIP

Bax, Ad, 1956–
Two-dimensional nuclear magnetic resonance in liquids.

Includes bibliographical references.
1. Liquids—Optical properties. 2. Liquids—Magnetic properties.
3. Nuclear magnetic resonance spectroscopy. I. Title.
QC145.4.06B38 538'.362 82-627
ISBN 90-277-1412-6 AACR2

Published by Delft University Press, Mijnbouwplein 11, 2628 RT Delft, Holland and by
D. Reidel Publishing Company, P.O. Box 17, 3300 AA Dordrecht, Holland

Sold and distributed in the U.S.A. and Canada
by Kluwer Boston Inc.,
190 Old Derby Street, Hingham, MA 02043, U.S.A.

In all other countries, sold and distributed
by Kluwer Academic Publishers Group,
P.O. Box 322, 3300 AH Dordrecht, Holland

D. Reidel Publishing Company is a member of the Kluwer Group

Printed in The Netherlands

TO INGRID

Preface

The introduction by Jeener of the concept of two-dimensional (2D) Fourier transformation in NMR has initiated the development of a tremendous number of new experiments. Applications of the new methods range from investigation of organic and biochemical type molecules in solution to solid state samples and even spin imaging of the human body. Literature describing the new 2D NMR experiments is very fragmentary and little comprehensive reviews have appeared to date. Confusion among the users of the sometimes rather intricate and complicated experiments is therefore common and understandable.

In this study I present a description of those two-dimensional experiments of which I feel they are important for liquid state NMR studies of organic and biochemical compounds. However, the development of the field of NMR techniques is so rapid, that any attempt to write a complete study will be outdated by the time it appears. Nevertheless I hope that this work will be of use for those who want to get acquainted with 2D NMR and for those who wish to know some of the fine details of theory and experimental realization. The attention given to a certain experiment is not proportional to its importance. Especially the important experiment for the detection of cross-relaxation is underrated in his respect.

This book is essentially a modified copy of my doctoral thesis which was written with the intention of being published. Therefore, I am particularly indebted to my research advisors, Prof. Dr. J. Smidt and Dr. A.F. Mehlkopf from Delft University of Technolgy, without whom none of this work would have taken place. I am very much obliged to Dr. R. Freeman from Oxford University who invited me to stay as a guest in his research group. I am indebted to him and all the members of his group, in particular Dr. G.A. Morris, T.A. Frenkiel, S.P. Kempsell, M. Levitt and T.H. Mareci, for their hospitality and for continuous help in theoretical and experimental problems. I also gratefully acknowledge staff members at the Delft group for their instrumental and theoretical assistance, and last but not least, Ingrid Pufahl for typing the manuscript.

Ad Bax

CONTENTS

Drawings: A.S.G. de Knegt

This research was performed in the Magnetic Resonance Group of the Department of Technical Physics, Technische Hogeschool, Delft, and in Dr. R. Freeman's group at the Physical Chemistry Laboratory, University of Oxford. The investigation was supported by a grant from the Delft Hogeschool Fonds.

Concise list of symbols

		unit
$A(\omega)$	absorption line shape function	-
AT_1, AT_2	acquisition time along the t_1- and t_2-axis	s
C	constant	-
c_m	coefficient	-
$c_m^\times$	complex conjugated of c_m	-
$D(\omega)$	dispersion line shape function	-
E_k	energy of state $\|k\rangle$	J
$\hat{F}_{x,y,z}$	total spin angular momentum operator in the x,y,z dimension	rad^{-1}
F_1, F_2	frequency dimensions corresponding to t_1 and t_2	s^{-1}
$g(t)$	filtering function	-
$\hat{H}$	Hamiltonian	J
H_o	static magnetic field strength	Am^{-1}
H_1	strength of the irradiated radiofrequency field during a pulse	Am^{-1}
$\hbar$	$h/2\pi$, with h being Planck's constant ($h=6.6262\ 10^{-34}$ Js)	Js
$\hat{I}_{x,y,z}$	spin angular momentum operator in the x,y,z dimension	rad^{-1}
J	coupling constant	s^{-1}
k	Boltzmann's constant ($k=1.3807\ 10^{-23}\ JK^{-1}$)	JK^{-1}
M_o	thermal equilibrium magnetization	$Wb\ m^{-2}$
M_{tr}	transverse magnetization	$Wb\ m^{-2}$
M_k	magnetic quantum number of state $\|k\rangle$	-
P_k	population of energy level k	-
P_N	noise power	-
$\hat{R}_{x,y,z}(\alpha)$	rotation operator for a pulse with flip angle α applied along the x,y or z-axis	-
R_o	amplitude	$Wb\ m^{-2}$
S	frequency domain signal	-
s	time domain signal	-
T	(1) temperature	K
	(2) time	s
T_1	longitudinal relaxation time	s
T_2	transverse relaxation time	s
$T_2^\star$	decay constant describing inhomogeneity and transverse relaxation	s
t_1	length of the evolution period	s
t_2	running time during the detection period	s

γ	magnetogyric ratio	$mA^{-1}s^{-1}$
δ	chemical shift	–
$\Delta_1,\Delta_2,\Delta_3$	delays	s
τ	delay	s
ϕ	phase angle	rad
ψ	phase angle	rad
Ψ	wave function	–
σ	density matrix	–
Ω	angular frequency	rad s^{-1}
Ω_R	angular frequency of the rotating frame	rad s^{-1}
ω_1,ω_2	frequency variables	rad s^{-1}

INTRODUCTION

In 1945 the groups of both Bloch (1) and Purcell (2) succeeded in detecting nuclear magnetic resonance (NMR) absorption in bulk matter. This energy absorption was observed by irradiating the sample with a radiofrequency field and varying the strength of the magnetic field. The combination of the magnetic field strength at resonance and the frequency irradiated gives information about the chemical environment of the nucleus considered, i.e. the chemical shielding of the nucleus by the surrounding electrons (3). When field homogeneity was improved, and hence resolution enhanced, a new feature in magnetic resonance appeared: the spin-spin coupling (4,5). This opened the opportunity to the development of an extension of the technique with double resonance spectroscopy (6). In double resonance one frequency is irradiated at the resonance frequency of a certain nucleus while another nucleus is observed by sweeping the frequency of the observe transmitter through resonance condition. In these experiments the scalar interaction between the nuclei is the basis for investigations about the connectivity (7) in the spin system and the relative signs of coupling constants (8). Investigations were mainly limited to the abundant ^{1}H nucleus which, due to its high magnetogyric ratio, offers best sensitivity. When Ernst and Anderson (9) introduced the application of Fourier transformation of pulse responses in NMR spectroscopy (1966), this improved the sensitivity significantly. The gain in sensitivity allowed the spectra of many nuclei to be recorded routinely, including the rare but important ^{13}C and the ^{15}N nuclei. Due to the extensive amount of arithmetic needed for performing the Fourier transformation, in the early seventies computers were integrated in NMR spectrometers. Another task of this computer was the (indirect) control of the spectrometer, enabling the execution of the growing number of sophisticated experiments as e.g. longitudinal and transverse relaxation measurements (10,11), DEFT(12), SEFT(13), Rapid Scan (14), and selective population transfer experiments (15). The double resonance experiments maintained their function, most commonly for the assignment of spectra of homonuclear (^{1}H) or heteronuclear (^{13}C) coupled spins. In these experiments a pulse response is acquired while a certain part of the spectrum is irradiated with a continuous monochromatic radiofrequency field. The resulting spectrum is a function of two frequency parameters; one frequency coordinate is obtained after Fourier transformation of the pulse response, and the other is determined by the frequency of the irradiating r.f.field. Only a Fourier transformation with respect to one time variable is performed in this case.

Jeener (16) was the first (1971) to propose the idea of two-dimensional Fourier transformation of an NMR signal obtained as a function of two time variables, yielding a spectrum which is a function of two frequency variables. Jeener proposed his experiment as an alternative for the homonuclear double resonance experiments. It was much later (1976) before a detailed and rather complex theoretical description of two-dimensional Fourier transform NMR was presented by Ernst and co-workers (17). They extended the applicability of the principle to the indirect detection of multiple quantum transitions, which was only partly possible with the conventional one-dimensional experiments (18,19,20). Their publication was rapidly followed by an avalanche of new experiments using the concept of two-dimensional Fourier transformation. The three main categories which can be distinguished are:

(a) Shift correlation spectroscopy

This class of experiments can be considered as an alternative for the double resonance experiments, correlating the shifts of coupled nuclei, exchanging nuclei or nuclei which have a cross-relaxation interaction.

(b) J-spectroscopy

In two-dimensional J-spectroscopy the two parameters determining the resonance positions in a spectrum, i.e. the chemical shift and the scalar coupling are separated.

(c) Multiple quantum spectroscopy

Two-dimensional Fourier transformation experiments facilitate the detection and enlarge the applicability of multiple quantum transitions.

In this book an attempt will be made to explain the theory of the most important two-dimensional experiments in an elementary way. A main goal is that it can be used by chemists as a practical guide for the use of two-dimensional spectroscopy. The description will be limited to experiments on substances in the isotropic liquid phase. A survey of the different two-dimensional experiments available to date, their applicability and their practical limitation will be presented. New contributions by the author will be integrated in this survey.

In the first chapter some general aspects of two-dimensional NMR will be discussed. Apart from a quantum mechanical treatment also a classical and illustrative description of a simple two-dimensional NMR experiment will be given. Furthermore, the technical aspects, concerning data-processing and presentation will be considered.

The three main categories mentioned above will be treated in the chapters 2, 3 and 4 respectively.

Principles cf all three categories will be used in chapter 5 where a set of new experiments is described, all concerning the detection of interaction between carbon-13 nuclei in samples with natural carbon-13 abundance.

In chapter 6 a short summary will be given of which two-dimensional methods can be of practical use in solving certain problems.

Until now no quantitative comparison of two-dimensional spectroscopy with the conventional double resonance experiments has been made. However, it is generally believed that both resolution in the second frequency dimension and sensitivity of the two-dimensional methods are generally higher. In the final chapter 7 a survey will be given of those spectrometer requirements which are more critical for two-dimensional spectroscopy than for performing sophisticated one-dimensional experiments. As the shift correlation experiments can be automated in such a way that hardly any understanding of the actual two-dimensional experiment is required by the operator, it can be expected that in future the double resonance experiments will almost completely be replaced by two-dimensional spectroscopy. Since the extra demands mainly apply to the computer part of the spectrometer, and the costs of this part are still decreasing, this probably will not stop a further development of two-dimensional NMR.

Since part of the research work which has led to this book has been performed in the group of Dr. R. Freeman in Oxford, many of the spectra shown in this book are recorded on a Varian XL 200 spectrometer, available in his laboratory. The other spectra are recorded on a home-built 7T HR-NMR spectrometer (21).

References

(1) F. Bloch, W.W. Hansen and M. Packard, Phys. Rev. 69, 127 (1946).

(2) E.M. Purcell, H.C. Torrey and R.V. Pound, Phys. Rev. 69, 37 (1946).

(3) J.T. Arnold, P. Dharmatti and M. Packard, J. Chem. Phys. 19, 507 (1951).

(4) E.L. Hahn, D.E. Maxwell, Phys. Rev. 88, 1070 (1952).

(5) H.S. Gutowsky, D.W. McCall and C.P. Slichter, J. Chem. Phys. 21, 279 (1953).

(6) W.A. Anderson and R. Freeman, J. Chem. Phys. 37, 85 (1962).

(7) W.A. Anderson, R. Freeman and C.A. Reilly, J. Chem. Phys. 39, 1518 (1963).

(8) R. Freeman and D.M. Whiffen, Mol. Phys. 4, 321 (1961).

(9) R.R. Ernst and W.A. Anderson, Rev. Sci. Instr. 37, 93 (1966).

(10) R.L. Vold, J.S. Waugh, M.P. Klein and D.E. Phelps, J. Chem. Phys. 48, 3831 (1968).

(11) H.Y. Carr and E.M. Purcell, Phys. Rev. 94, 630 (1954).

(12) E.D. Becker, J.A. Ferretti and T.C. Farrar, J. Am. Chem. Soc. 91, 7784 (1969).

(13) A. Allerhand and D.W. Cohran, J. Am. Chem. Soc. 92, 4482 (1970).

(14) J. Dadok and R.F. Spacher, J. Magn. Reson. 13, 243 (1974).

(15) S. Sorensen, R.S. Hansen and H.J. Jakobsen, J. Magn. Reson. 14, 243 (1974).

(16) J. Jeener, Ampere International Summer School, Basko Polje, Yugoslavia 1971.

(17) W.P. Aue, E. Bartholdi and R.R. Ernst, J. Chem. Phys. 64, 2229 (1976).

(18) W.A. Anderson, Phys. Rev. 104, 850 (1956).

(19) W.A. Anderson, R. Freeman and C.A. Reilly, J. Chem. Phys. 39, 1518 (1963).

(20) P. Bucci, M. Martinelli and S. Santucci, J. Chem. Phys. 52, 4041 (1970), J. Chem. Phys. 53, 4524 (1970).

(21) A.F. Mehlkopf, Thesis, Delft 1978.

1 GENERAL ASPECTS OF TWO-DIMENSIONAL NMR SPECTROSCOPY

1.1 Introduction

In this book a two-dimensional (2D) spectrum $S(\omega_1,\omega_2)$ is defined as a spectrum which is a function of two independent frequency variables ω_1 and ω_2, and is obtained after a two-dimensional Fourier transformation of a time function $s(t_1,t_2)$. The two frequency axes of such a spectrum will be labelled F_1 and F_2 instead of ω_1 and ω_2. As will be shown, the information content of such a 2D spectrum depends entirely on the experimental scheme used. However, as has been shown by Ernst and co- workers (1), it is possible to give a general quantum mechanical description, which includes almost all two-dimensional NMR experiments.

In this chapter a simplified version of this quantum mechanical description will be given, followed by an illustrative but less general classical approach, in combination with some experimental aspects of two-dimensional spectroscopy. Relaxation effects and magnetic field inhomogeneity will be neglected in this chapter unless explicitly stated.

In principle in all two-dimensional experiments four different time-intervals can be distinguished, (a) a preparation period, (b) an evolution period, (c) a mixing period (which may be absent) and (d) a detection period (Fig. 1.1). The signal is detected* as a function of the running time variable t_2, with the length of the evolution period t_1 as a parameter. By repeating the experiment for a large number of different times t_1, keeping all other set-

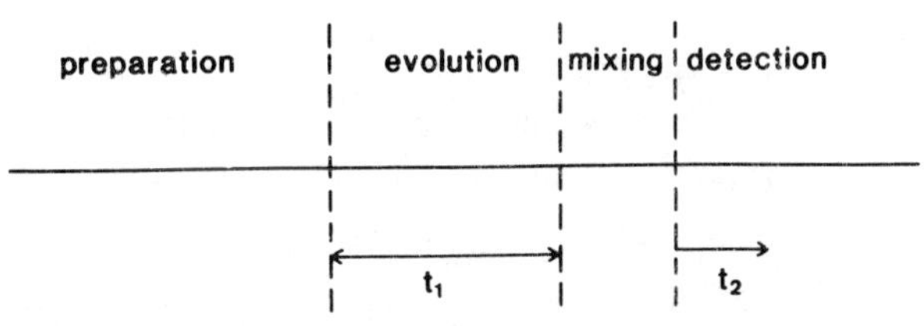

Fig.1.1 General subdivision of the time axis in a two-dimensional experiment. The preparation period usually consists of a long delay time, to reach thermal equilibrium , followed by one or more r.f.pulses at the end. During the evolution period the (non-equilibrium) spin system evolves. The delay time t_1 is varied in the successive experiments. The mixing period consists of pulses and delays with fixed lengths. During the detection period the signal is acquired as a function of t_2.

---

* In this study detection during the time t_2 is always assumed to be in the quadrature mode.

tings constant, a two-dimensional time signal $s(t_1,t_2)$ is obtained. In the corresponding two-dimensional frequency spectrum, the behaviour of the spin system during both evolution and detection periods is shown simultaneously. Often the correlation between these two behaviours is the useful extra source of information.

1.2 Quantum mechanical description

A simplified general quantum mechanical description of two-dimensional NMR spectroscopy is given below. The effects of relaxation are not taken into account. Since the treatment relies on the use of the density matrix (2) to describe the spin system, a short introduction of this density matrix will be given first.

1.2.1 Phase coherence and the density matrix

It is well-known that in an isolated spin system there are only a limited number of stationary wave functions, i.e. eigenstates. In a coupled spin system , consisting of N coupled spin 1/2 nuclei, there are 2^N different eigenstates. The eigenstates will be labelled $|1\rangle$, $|2\rangle$, etc., and the corresponding energies are E_1, E_2, etc. The state Ψ of a certain isolated spin system can always be described as a linear combination of eigenstates:

$$\Psi = \sum_{n=1}^{N} c_n |n\rangle \; . \qquad [1.1]$$

The expectation value $\langle A\rangle$ of an operator $\hat{A}$ is given by

$$\langle A\rangle = \langle\Psi|\hat{A}|\Psi\rangle = \sum_{n=1}^{N}\sum_{m=1}^{N} c_n c_m^{\times}\langle m|\hat{A}|n\rangle \; , \qquad [1.2]$$

in which $\langle m|\hat{A}|n\rangle$ is a constant A_{mn} which is independent of the particular state Ψ of the system. In the case of a macroscopic sample, the eigenfunctions are the same over the whole sample, but the coefficients c_n may vary, and hence the expectation value will be ensemble average of the expectation values of operator $\hat{A}$ for all individual spin systems:

$$\langle\overline{A}\rangle = \overline{\langle\Psi|\hat{A}|\Psi\rangle} = \sum_{n=1}^{N}\sum_{m=1}^{N} \overline{c_m^{\times}c_n} A_{mn} \; . \qquad [1.3]$$

The quantities $\overline{c_m^{\times}c_n}$ can be considered as matrix elements σ_{nm} of a matrix σ, and it is this matrix that is called the density matrix. The coefficients c_n are complex and can be written as

$$c_n = |c_n|\exp(i\alpha_n). \qquad [1.4]$$

Using the assumption that c_n and α_n are statistically independent, it then follows that density matrix element σ_{nm} can be written as

$$\sigma_{nm} = \overline{c_m^{\times} c_n} = \overline{|c_n||c_m|}\ \overline{\exp\{i(\alpha_n - \alpha_m)\}}. \qquad [1.5]$$

An element σ_{nm} will have a value not equal to zero if the product $\overline{|c_n||c_m|}$ is not equal to zero and if the average value of $\overline{\exp\{i(\alpha_n - \alpha_m)\}}$ is not zero, i.e. if there is a coherence between the phases α_m and α_n. Therefore σ_{nm} describes the phase coherence between the states $|n\rangle$ and $|m\rangle$. If in the states $|m\rangle$ and $|n\rangle$ only one spin differs in polarization, by one unit, such a coherence corresponds, due to the selection rules (2 p.13) , to physically observable transverse magnetization. A coherence between two states $|m\rangle$ and $|n\rangle$ in which more than one spin differs in polarization, or one spin differs by more than one unit, is called a multiple quantum coherence, in literature often also referred to as multiple quantum transition. A diagonal matrix element σ_{mm} in a normalized density matrix ($\sum_{n=1}^{N} \sigma_{nn} = 1$) is equal to the relative population of level m. As follows immediately from Eq. [1.5] , the density matrix is always Hermitian, i.e. $\sigma_{mn} = \sigma_{nm}^{\times}$.

The time dependence of a state Ψ in a physical environment described by the time-independent Hamiltonian $\hat{H}$ follows from the Schrödinger equation:

$$-\frac{\hbar}{i}\frac{d\Psi}{dt} = \hat{H}\,\Psi. \qquad [1.6]$$

Substitution of Eq. [1.1] and solving the differential equation gives

$$\Psi(t) = \sum_{n=1}^{N} c_n(0)\ \exp(-iE_n t/\hbar)\ |n\rangle\ , \qquad [1.7]$$

where $c_n(0)$ denotes the coefficient c_n at time zero. Using the assumption that $\hat{H}$ is constant over the whole sample, the time dependence of σ_{nm} is derived from Eqs. [1.4] , [1.5] and [1.7]:

$$\sigma_{nm}(t) = \exp\{i(E_m - E_n)t/\hbar\}\ \sigma_{nm}(0)\ , \qquad [1.8]$$

where $\sigma_{nm}(0)$ is the density matrix element at time zero. More general the dynamic behaviour of the density matrix is described by

$$\sigma(t) = \exp(-i\hat{H}t/\hbar)\sigma(0)\exp(i\hat{H}t/\hbar), \qquad [1.9]$$

assuming again that the Hamiltonian $\hat{H}$ is time-independent.

All derivations given so far are based on a description in a laboratory frame. Since in practice a radiofrequency field of frequency $\Omega_R/2\pi$ and a lock in detector with the same reference frequency are used to observe the spin system, it is useful to transform all equations to a frame, rotating with frequency $\Omega_R/2\pi$ about the z-axis, which is chosen to be parallel to the magnetic field. In the Schrödinger representation this can be done by replacing any operator $\hat{A}$ by an operator $\hat{A}_R$ (3):

$$\hat{A}_R = \exp(i\Omega_R\hat{F}_z t)\hat{A}\exp(-i\Omega_R\hat{F}_z t), \tag{1.10}$$

where $\hat{F}_z$ is the total spin angular momentum operator along the z-axis:

$$\hat{F}_z = \sum_{n=1}^{N} \hat{I}_{z_n} . \tag{1.11}$$

The density matrix in the rotating frame is then given by

$$\sigma_R(t) = \exp(i\Omega_R\hat{F}_z t)\sigma(t)\exp(-i\Omega_R\hat{F}_z t). \tag{1.12}$$

The development through time of a density matrix element $\sigma_{R_{mn}}(t)$ is, analogous to Eq. [1.8] described by

$$\sigma_{R_{mn}}(t) = \exp\{i(m_m - m_n)\Omega_R t\}\exp(i\omega_{mn}t)\sigma_{R_{mn}}(0), \tag{1.13}$$

where $\omega_{mn} = (E_m - E_n)/\hbar$ and m_n denotes the total magnetic quantum number of state $|n\rangle$. All discussions below, will be in the rotating frame and therefore the index 'R' will be omitted.

1.2.2 The effect of a radiofrequency pulse

The effect of a pulse with flip angle α, applied along the x-axis of the rotating frame can be calculated using Eq. [1.9], and is described by a rotation operator $\hat{R}_x(\alpha)$, which relates the density matrix σ' after this pulse to the old one, according to

$$\sigma' = \hat{R}_x^{-1}(\alpha)\sigma\,\hat{R}_x(\alpha), \tag{1.14}$$

where $\hat{R}_x(\alpha)$ can be calculated from

$$\hat{R}_x(\alpha) = \exp(-i\alpha\hat{F}_x);\hat{R}_x^{-1}(\alpha) = \exp(i\alpha\hat{F}_x) \tag{1.15}$$

and

$$\hat{F}_x = \sum_{n=1}^{N} \hat{I}_{x_n} . \tag{1.16}$$

A similar rotation operator $\hat{R}_y(\alpha)$ can be used to describe the effect of a pulse applied along the y-axis.

For a weakly coupled N spin-$\frac{1}{2}$ system, explicit general expressions for

$\hat{R}_x(\alpha)$ and $\hat{R}_y(\alpha)$ have been derived by Schaüblin et al. (4).

1.2.3 The two-dimensional experiments

In principle the tools given in the previous two sections, are sufficient to calculate the evolution of the spin system during any kind of pulse sequence, provided that expressions for the Hamiltonian $\hat{H}$ and the rotation operators $\hat{R}$ are available.

In this section it will be pointed out how to calculate the relation between the initial thermal equilibrium spin system and the signals detected in a two-dimensional experiment. For reasons of simplicity relaxation will be neglected and a time-independent physical environment (apart from applied r.f.pulses) will be assumed.

As pointed out in section 1.1, the two-dimensional experiment consists basically of four periods: the preparation period, the evolution period, the mixing period and the detection period. The following notations for the density matrix will be used to describe the spin system at a certain time:

σ_{eq} - thermal equilibrium
$\sigma(0)$ - at the end of the preparation period
$\sigma(t_1)$ - at the end of the evolution period
$\sigma(t_1,0)$ - at the end of the mixing period
$\sigma(t_1,t_2)$ - at time t_2 during the detection period.

Below, the relations between σ_{eq}, $\sigma(0)$, $\sigma(t_1)$, $\sigma(t_1,0)$ and $\sigma(t_1,t_2)$ will be discussed.

The preparation period

The first part of the preparation period usually consists of a delay time long compared with the longitudinal relaxation time of the nuclear spins. At the end of the preparation period, usually a non-equilibrium spin system is created by means of a pulse sequence. Which pulse sequence is used, depends on the type of experiment.

A few examples are given below:

(1) Single non-selective★ $(\pi/2)_x$ pulse

★ The following names will be used in this study to indicate different types of pulses:
line-selective pulse - pulse affecting only the magnetization which corresponds to a certain resonance line.
multiplet-selective pulse - pulse affecting all magnetization components corresponding to resonance lines in a certain multiplet in an identical way. Other magnetizations are unaffected.
nucleus-selective pulse - pulse affecting all magnetization components corresponding to a certain type of nuclei (e.g. protons) in an identical way.
non-selective pulse - pulse affecting all magnetization components in an identical way.

A single non-selective $(\pi/2)_x$ pulse at the end of the preparation period is used in for example homonuclear shift correlation spectroscopy (section 2.3) and homonuclear J-spectroscopy (section 3.3). In this case the relation between σ_{eq} and $\sigma(0)$ is found by using Eqs.[1.14], [1.15] and [1.16]:

$$\sigma(0) = \exp(i\pi/2\hat{F}_x)\sigma_{eq}\exp(-i\pi/2\hat{F}_x). \quad [1.17]$$

(2) Single nucleus-selective $(\pi/2)_x$ pulse

A single selective $(\pi/2)_x$ pulse, affecting only one nucleus (or type of nuclei) is used at the end of the preparation period in for example heteronuclear shift correlation spectroscopy(section 2.3) and heteronuclear J-spectroscopy (section 3.2). The relation between $\sigma(0)$ and σ_{eq} is analogous to Eq.[1.17] given by

$$\sigma(0) = \exp(i\pi/2\hat{I}_x)\sigma_{eq}\exp(-i\pi/2\hat{I}_x), \quad [1.18]$$

where $\hat{I}_x$ is the spin angular momentum operator of the irradiated nucleus.

(3) Non-selective $(\pi/2)_x - \tau - (\pi/2)_x$ sequence

In many multiple quantum experiments (chapter 4) two non-selective $(\pi/2)_x$ pulses spaced by a time delay τ are used at the end of the preparation period.$\sigma(0)$ can then be calculated by using Eqs.[1.9] and [1.14]:

$$\sigma(0) = \exp(i\pi/2\hat{F}_x)\exp(-i\hat{H}\tau/\hbar)\exp(i\pi/2\hat{F}_x)\sigma_{eq}\exp(-i\pi/2\hat{F}_x) \times \times \exp(i\hat{H}\tau/\hbar)\exp(-i\pi/2\hat{F}_x). \quad [1.19]$$

If this sequence is applied to a coupled spin system, explicit calculation shows that $\sigma(0)$ generally contains matrix elements $\sigma_{mn}(0) \neq 0$, even if more than one spin differs in polarization in the states m and n, thus describing multiple quantum coherence.

The evolution period

During the evolution period the non-equilibrium spin system evolves under influence of a physical environment described by Hamiltonian $\hat{H}$. The relation between $\sigma(0)$ and $\sigma(t_1)$ is then given by Eq.[1.9]:

$$\sigma(t_1) = \exp(-i\hat{H}t_1/\hbar)\sigma(0)\exp(i\hat{H}t_1/\hbar). \quad [1.20]$$

However, in many experiments {e.g. heteronuclear shift correlation (section 2.3) and J-spectroscopy (chapter 3) the evolution is interrupted by the application of a (nucleus)-selective π pulse in the centre of the evolution period.

In this case $\sigma(t_1)$ is given by

$$\sigma(t_1) = \exp(-i\,\hat{H}t_1/2\hbar)\exp(i\pi\hat{F}_x)\exp(-i\,\hat{H}t_1/2\hbar)\sigma(0)\exp(i\,\hat{H}t_1/2\hbar) \times \times \exp(-i\pi\hat{F}_x)\exp(i\,\hat{H}t_1/2\hbar). \quad [1.21]$$

Usually a cumbersome explicit calculation, using Eq. [1.21] can be avoided by defining a new Hamiltonian $\bar{\hat{H}}$, which describes the average effect of the physival environment over the total evolution period t_1, including the effect of the π pulse. In many cases the form of $\bar{\hat{H}}$ is rather straight forward, e.g. in homonuclear J-spectroscopy (section 3.3) it is found by taking out the terms in $\hat{H}$ which describe the interaction with the static magnetic field (only permitted in the case of weak coupling).

For example in the case of a homonuclear coupled AX spin system one finds

$$\hat{H} = (2\pi)^{-1}\gamma H_0\{\hat{I}_{zA}(1-\delta_A) + \hat{I}_{zX}(1-\delta_X)\} + J\hat{I}_A\cdot\hat{I}_X\ , \quad [1.22a]$$

$$\bar{\hat{H}} = J\hat{I}_A\cdot\hat{I}_X, \quad [1.22b]$$

where γ is the magnetogyric ratio, δ_A the chemical shift of nucleus A, H_0 the strength of the magnetic field, and J the coupling constant.

The mixing period

The mixing period usually consists of a single non-selective pulse {e.g homonuclear shift correlation (section 2.3) and homonuclear multiple quantum spectroscopy (chapter 5)} or a set of nucleus-selective pulses {e.g. heteronuclear shift correlation (section 2.2)}, possibly preceded and followed by delays Δ_1 and Δ_2. Generally the mixing period is thus often of the form (Δ_1-pulse(s)-Δ_2). In order to obtain some explicit results, the delays Δ_1 and Δ_2 will not be taken into account in the following discussion, and a mixing period consisting of only one pulse or a set of pulses is assumed. The density matrix after the mixing pulses $\sigma(t_1,0)$ is related to $\sigma(t_1)$ by:

$$\sigma(t_1,0) = \hat{R}^{-1}\sigma(t_1)\hat{R} \quad [1.23a]$$

with

$$\hat{R} = \prod_{n=1}^{N} \exp(-i\theta_n\hat{I}_{xn}). \quad [1.23b]$$

In Eq. [1.23b] it is assumed that nucleus n feels a pulse with flip angle θ_n applied along the x-axis. For a spin system of N non-equivalent weakly coupled spin 1/2 nuclei the rotation operator $\hat{R}$ can be represented by a $2^N \times 2^N$ matrix, with elements R_{kl}. Using this concept, the transfer due to the pulse(s), from a density matrix element $\sigma_{mn}(t_1)$ to a matrix element $\sigma_{kl}(t_1, 0)$, denoted by $\sigma_{kl,mn}(t_1,0)$ is given by

$$\sigma_{kl,mn}(t_1,0) = (R^{-1})_{km}R_{nl}\sigma_{mn}(t_1). \qquad [1.24]$$

The detection period

During the detection period, the density matrix at time t_2 can be calculated using [1.9]:

$$\sigma(t_1,t_2) = \exp(-i\hat{H}t_2/\hbar)\sigma(t_1,0)\exp(i\hat{H}t_2/\hbar). \qquad [1.25]$$

The detected transverse magnetization is then given by

$$M_{tr}(t_1,t_2) = C\ \mathrm{Tr}(\hat{F}_x + i\hat{F}_y)\sigma(t_1,t_2), \qquad [1.26]$$

where C is a constant proportional to the magnetization in thermal equilibrium (M_o).

For the detected transverse magnetization $M_{tr(kl),(mn)}$ which corresponds to coherence kl and which originates from coherence mn during the evolution period one finds from Eqs. [1.13], [1.20], [1.24], [1.25] and [1.26]

$$M_{tr(kl),(mn)}(t_1,t_2) = C\{Z_{kl,mn}\exp(i\omega_{mn}t_1)\exp(i\omega_{kl}t_2) + Z_{kl,nm}\exp(-i\omega_{mn}t_1)\exp(i\omega_{kl}t_2) \qquad [1.27]$$

with $Z_{kl,mn}$ defined by

$$Z_{kl,mn} = 2\mathrm{Re}(F_{x1k} + iF_{y1k})(R^{-1})_{km}R_{nl}\sigma_{mn}(0).$$

The total detected signal is given by

$$M_{tr}(t_1,t_2) = \sum_{k=1}^{2^N}\sum_{l=k}^{2^N}\sum_{m=1}^{2^N}\sum_{n=m}^{2^N} M_{tr(kl),(mn)}(t_1,t_2). \qquad [1.28]$$

In the case of a mixing pulse with flip angle $\pi/2$, or a set of multiplet-selective mixing pulses each with flip angle $\pi/2$, or if $m \neq k,l$ and $n \neq k,l$ it can be derived (1) that

$$|Z_{kl,mn}| = |Z_{kl,nm}|. \qquad [1.29]$$

Eq. [1.27] then can be simplified to the form

$$M_{tr(kl),(mn)}(t_1,t_2) = 2C\ Z_{kl,mn}\cos(\omega_{mn}t_1 + \phi_{kl,mn})\exp(i\omega_{kl}t_2), \qquad [1.30]$$

where $\phi_{kl,mn}$ is a constant.

Eq.[1.30] shows that the part of a detected magnetization which corresponds with coherence kl and originates from coherence mn is modulated in amplitude as a function of t_1, with the frequency ω_{mn}.

A complex two-dimensional Fourier transformation of signals of the form of Eq.[1.27] will give two resonance lines, one at $(\omega_1,\omega_2) = (\omega_{mn},\omega_{kl})$, provided that $Z_{kl,mn} \neq 0$ and one at $(-\omega_{mn},\omega_{kl})$ provided that $Z_{kl,nm} \neq 0$. If m = k and n = l the peaks will fall on the diagonals $\omega_1 = \pm\omega_2$, and are referred to as diagonal peaks (Fig.1.2). In the case where mn and kl are different coherences, with generally different frequencies, the corresponding peaks are referred to as cross-peaks. In the case where m = n and thus $\omega_{mn} = 0$, the peaks fall on the line $\omega_1 = 0$ and are referred to as axial peaks. These peaks represent the transfer from diagonal density matrix elements (corresponding to longitudinal magnetization) into observable transverse magnetization. These peaks thus contain information about the longitudinal relaxation process during the time t_1, but they are generally only of little interest, and can obscure the presence of other, much smaller resonances in the spectrum, because they can be of high intensity and can have long tails. Therefore, one usually tries to eliminate these peaks.

In the case of pure amplitude modulation (Eq.[1.30]) the spectrum will be

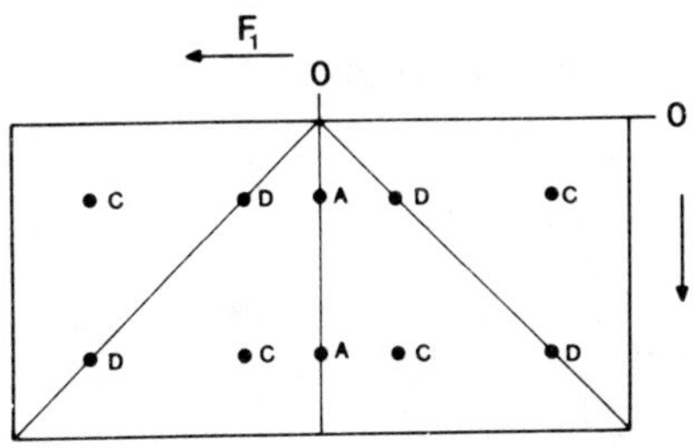

Fig.1.2 Schematic diagram of a two-dimensional spectrum. Axial peaks are indicated with 'A', diagonal peaks with 'D' and cross peaks with 'C'. Since the positions of the peaks are symmetrical about the line $F_1 = 0$, usually only the left or the right half of the spectrum is calculated and displayed.

symmetrical about the line $F_1 = 0$. In that case a real Fourier transformation with respect to t_1 should be used (see section 1.5.2), yielding only the right half of the spectrum of Fig.1.2, with the left (identical) half folded on top of it.

A practical example of a density matrix calculation using the theory described above is given in Appendix I.

1.2.4 Restrictions on magnetization transfer

As follows from Eq.[1.27] it is generally possible to transfer part of a coherence or the total coherence present between levels m and n to coherence between levels k and l, provided that the rotation matrix elements $(R^{-1})_{km}$ and R_{nl} and/or the elements $(R^{-1})_{kn}$ and R_{ml} are not zero. If both coherences mn and kl are single quantum coherences, this means that magnetization is transferred from one coherence to the other. However, there appears to be a restriction on the transfer of magnetization from one nucleus to another: As will be shown below the magnitude of the total magnetization of a nucleus has to be unchanged just before and just after a (set of) mixing pulse(s), provided that the pulses can all be considered as non-selective, nucleus-selective or multiplet-selective (see footnote section 1.2.3).

Consider again a system of N weakly coupled spins, described by a density matrix σ. The magnetization corresponding to a nucleus m is given by

$$M_{\alpha_m} = \mathrm{Tr}(\hat{I}_{\alpha_m}\sigma). \qquad \alpha = x,y,z \qquad [1.31]$$

A general set of mixing pulses is described by a rotation operator $\hat{R}$ of the form of Eq.[1.23] and after these mixing pulses the density matrix σ' is given by

$$\sigma' = \{\prod_{n=1}^{N} \exp(i\theta_n \hat{I}_{x_n})\}\, \sigma \{\prod_{n=1}^{N} \exp(-i\theta_n \hat{I}_{x_n})\}. \qquad [1.32]$$

For the magnetization of nucleus m after the pulses one finds by substitution:

$$\begin{aligned} M'_{\alpha_m} &= \mathrm{Tr}[\hat{I}_{\alpha_m}\{\prod_{n=1}^{N} \exp(i\theta_n \hat{I}_{x_n})\}\sigma\{\prod_{n=1}^{N} \exp(-i\theta_n \hat{I}_{x_n})\}] = \qquad [1.33] \\ &= \mathrm{Tr}[\{\prod_{n=1}^{N} \exp(-i\theta_n \hat{I}_{x_n})\}\hat{I}_{\alpha_m}\{\prod_{n=1}^{N} \exp(i\theta_n \hat{I}_{x_n})\}\sigma] = \\ &= \mathrm{Tr}[\exp(-i\theta_m \hat{I}_{x_m})\hat{I}_{\alpha_m}\exp(i\theta_m \hat{I}_{x_m})\sigma] \\ &\qquad \alpha = x,y,z \end{aligned}$$

since $[\hat{I}_{x_m}, \hat{I}_{\alpha_n}] = 0$ for $n \neq m$.

Using

$$[\hat{I}_{x_m},\hat{I}_{y_m}] = i\hat{I}_{z_m}, \qquad [1.34]$$

and cyclic permutations, one finds from Eq.[1.33]

$$\begin{aligned} M'_{x_m} &= M_{x_m} \\ M'_{y_m} &= \cos(\theta_m)M_{y_m} + \sin(\theta_m)M_{z_m} \\ M'_{z_m} &= \cos(\theta_m)M_{z_m} - \sin(\theta_m)M_{y_m} . \end{aligned} \qquad [1.35]$$

From Eq.[1.35] it follows that the total magnitude of the magnetization of nucleus m, $(M_{x_m}^2 + M_{y_m}^2 + M_{z_m}^2)^{\frac{1}{2}}$, just before and just after the mixing pulse is unchanged, i.e. no net magnetization is transferred from one nucleus to another. This means that the transferred multiplet components are in anti-phase, just after the mixing pulse. However, a time t_2 later, after they have precessed with different frequencies, net transferred magnetization can be present.

1.3 Coherence transfer echoes

An important phenomenon in many two-dimensional NMR experiments is the existence of coherence transfer echoes. A detailed theoretical description of the coherence transfer echo has been given by Maudsley et al. (5). In this section only a short and simple mathematically oriented explanation will be given. All frequencies used in this section are frequencies in the laboratory frame. As will be shown below there appears to be a coherence transfer echo effect in all two-dimensional NMR experiments where the magnitude of a detected coherence kl is modulated as a function of t_1. In this case the detected transverse magnetization can analogous to Eq.[1.30] be written as

$$M_{tr(kl),(mn)}(t_1,t_2) = C\cos(\omega_{mn}t_1 + \phi)\exp(i\omega_{kl}t_2). \qquad [1.36]$$

Neglecting the phase constant ϕ, Eq.[1.36] can be rewritten as

$$\begin{aligned} M_{tr(kl),(mn)}(t_1,t_2) = \tfrac{1}{2}C[&\exp\{i(-\omega_{mn}t_1 + \omega_{kl}t_2)\} + \\ &+ \exp\{i(\omega_{mn}t_1 + \omega_{kl}t_2)\}] . \end{aligned} \qquad [1.37]$$

Thus the signal detected during t_2 can be considered as the sum of two magnetization components which have phases at $t_2 = 0$, modulated in opposite sense as a function of t_1. Apart from the effect of scalar interaction both ω_{mn} and ω_{kl} are linearly dependent on the magnetic field strength. This implies that the frequency ratio ω_{mn}/ω_{kl} is independent of the magnetic field strength. From substitution of

$$t'_2 = t_1(\omega_{mn}/\omega_{kl}) \qquad [1.38]$$

in Eq.[1.37], it can be seen that at time t'_2 the phase of the signal detected, corresponding to the first part of Eq.[1.37] will be independent of the magnetic field strength to first order. In the case of magnetic field inhomogeneity the magnetic field strength varies for different parts of the sample, but at the time t'_2, given by Eq.[1.38], magnetization components $M_{tr(kl),(mn)}$ have identical phase in the whole sample, as far as the first term at the right hand side of Eq.[1.37] is concerned. This means that a so-called coherence transfer echo is formed. The magnetization corresponding to the second term of Eq.[1.37] which does not refocus during t_2 is generally referred to as antiecho.

Note that if for example magnetization is transferred from a single quantum proton transition to a carbon-13 transition, it follows immediately from Eq.[1.38] that the centre of the echo occurs at $t_2 = 4t_1$. The echo and antiecho signals can be separately detected by using suitable phase cycling (6,7,8) or by using pulsedfield gradients (9). Distinguishing between echo and antiecho signals is of importance in hetero- and homonuclear correlation spectroscopy (sections 2.2.3 and 2.3.2) and in multiple quantum spectroscopy (sections 4.4.2 and 4.4.3).

The definition of a coherence transfer echo given in this section is wider than in the original paper by Maudsley et al. (5), since now cases in which coherence is transferred from one multiplet component to another one within the same multiplet are also included.

1.4 Classical description of a simple two-dimensional experiment

The quantum mechanical treatment given in section 1.2 is sufficient to make a detailed analysis of most two-dimensional experiments, but it gives little insight into what is happening physically. In this section an example of a 2D experiment will be given, which can be completely described classically, using the Bloch equations (10).

A set of isolated (non-coupled) nuclear spins in a homogeneous magnetic field with strength H_o is considered. In this example the preparation period consists of a time long compared with the longitudinal relaxation time T_1 of the spins. So, at the end of the preparation period, a magnetization will exist, with magnitude M_o, pointing parallel to the magnetic field. After the preparation period, during a time t_1 a r.f.field with strength H_1 is switched on along the x-axis of a frame rotating with frequency Ω_R close to the resonance frequency γH_o of the nuclear spins (Fig.1.3a). Neglecting off-resonance effects, the magnetization will rotate during t_1 with angular frequency $\Omega_1 = \gamma H_1$ about the x-axis in the yz plane (Fig.1.3b). After switching off the r.f.field the magnetization will precess with angular frequency $\Omega_2 = \gamma H_o - \Omega_R$ about the z-axis in the rotating frame. Using quadrature detection★ the signal will then be given by

$$s(t_1,t_2) = M_o \sin(\Omega_1 t_1)\exp(i\Omega_2 t_2)\exp(-t_1/T_2^{(1)})\exp(-t_2/T_2^{(2)}). \qquad [1.39]$$

The decay of the magnetization during evolution and detection periods is expressed by the decay constants $T_2^{(1)}$ and $T_2^{(2)}$ respectively. Since in practice the decay rate $T_2^{(1)}$ will be very short due to inhomogeneity of the H_1-field, longitudinal relaxation during the evolution period can be neglected.

Fourier transformation of the response obtained for a certain value of t_1 gives

$$S(t_1,\omega_2) = M_o \sin(\Omega_1 t_1)\exp(-t_1/T_2^{(1)})\{A_2(\omega_2) + iD_2(\omega_2)\} \qquad [1.40]$$

with

$$A_2(\omega_2) = \frac{T_2^{(2)}}{1 + T_2^{(2)^2}(\Omega_2-\omega_2)^2}, \quad D_2(\omega_2) = \frac{T_2^{(2)^2}(\Omega_2-\omega_2)}{1 + T_2^{(2)^2}(\Omega_2-\omega_2)^2},$$

where A and D indicate absorption and dispersion respectively.

Fig.1.4a shows a set of spectra obtained this way for different values of t_1.

Only the real part of $S(t_1,\omega_2)$ is shown.

A cross-section parallel to the t_1-axis for $\omega_2 = \Omega_2$ is then described by

$$S(t_1,\Omega_2) = M_o T_2^{(2)} \sin(\Omega_1 t_1)\exp(-t_1/T_2^{(1)}). \qquad [1.41]$$

Real Fourier transformation of this cross-section with respect to t_1, and only considering positive frequencies, will then yield a resonance line

---

★ In quadrature detection the x- and the y-components M_x and M_y of the transverse magnetization M_{tr}, which makes an angle $\Omega_2 t_2$ with the y-axis, are detected simultaneously. This gives for the two components along the y- and x-axis: $M_y = M_{tr}\cos(\Omega_2 t_2)$ and $M_x = M_{tr}\sin(\Omega_2 t_2)$. Hence, the complex quadrature signal $s(t_2)$ can be written as: $s(t_2) = M_y + iM_x = M_{tr}\exp(i\Omega_2 t_2)$.

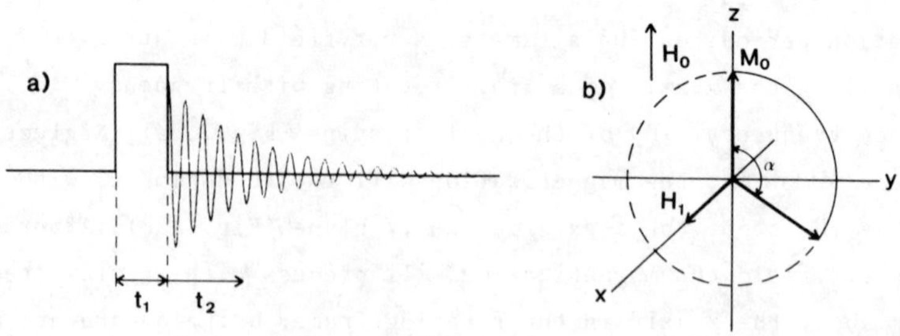

Fig.1.3 (a) The scheme of the two-dimensional experiment as discussed in the text. The preparation period consists of a long delay only. During the evolution period a strong r.f.field H_1 is applied. The mixing period is absent in this experiment.
(b) The precession with frequency γH_1 during the evolution period of the magnetization M_0 about the x-axis.

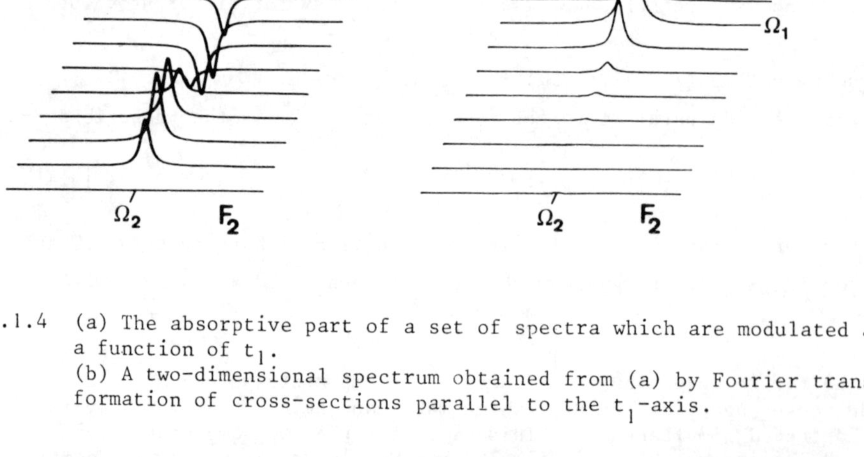

Fig.1.4 (a) The absorptive part of a set of spectra which are modulated as a function of t_1.
(b) A two-dimensional spectrum obtained from (a) by Fourier transformation of cross-sections parallel to the t_1-axis.

at $\omega_1 = \Omega_1$:

$$S(\omega_1,\Omega_2) = M_o T_2^{(2)} i\{A_1(\omega_1) - iD_1(\omega_1)\}, \quad [1.42]$$

where $A_1(\omega_1)$ and $D_1(\omega_1)$ are defined in a similar way as $A_2(\omega_2)$ and $D_2(\omega_2)$ but with the indices '2' replaced by '1'.

Fourier transformation of all the cross-sections through the real part of $S(t_1,\omega_2)$ for different values of ω_2, parallel to the t_1-axis, gives

$$S(\omega_1,\omega_2) = M_o\{iA_2(\omega_2)A_1(\omega_1) + A_2(\omega_2)D_1(\omega_1)\}. \quad [1.43]$$

The imaginary part of this signal represents a double absorption Lorentzian line with its centre at $(\omega_1,\omega_2) = (\Omega_1,\Omega_2)$ (Fig.1.4b).

Practical use of this experiment can be made for mapping the magnitude of the r.f.field H_1 as a function of the position in the sample. If a strong linear field gradient G is applied along one of the axes, e.g. the z-axis, the sample can be divided into a large number of layers perpendicular to the z-axis, each with a magnetic field strength linearly dependent on z and resonance frequency

$$\Omega_2(z) = \Omega_2(0) + Gz \quad . \quad [1.44]$$

Each layer will give rise to a two-dimensional absorption line, but at an ω_2-frequency proportional to z, and, neglecting off-resonance effects, at an ω_1 frequency determined by the magnitude of the H_1-field in the layer considered. The magnitude of the r.f.field can thus be mapped as a function of position.

As an illustration in Fig.1.5 the r.f.field strength in a home-built probe for 300 MHz is shown as a function of position in the sample.

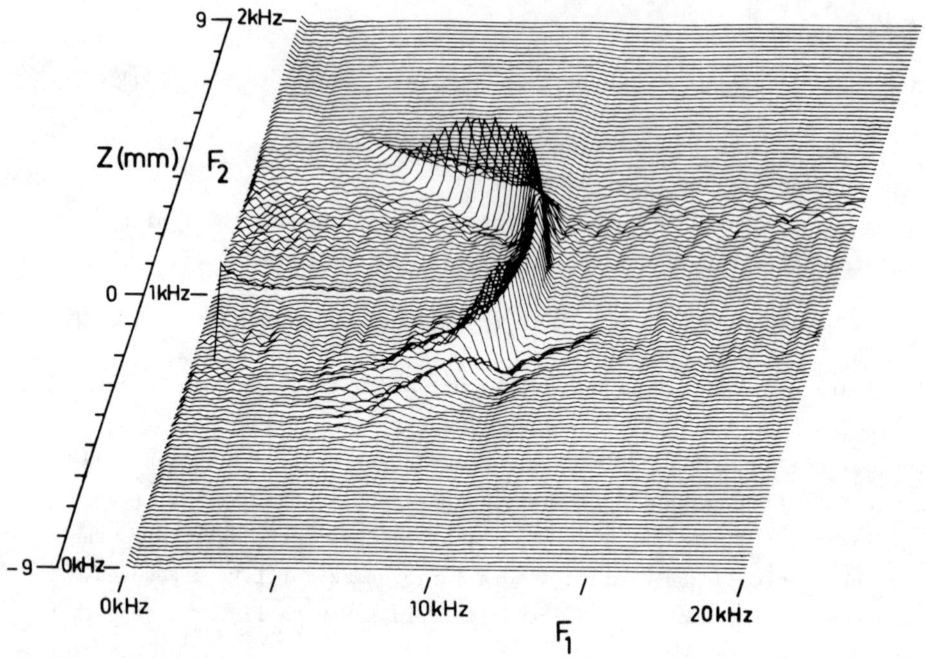

Fig.1.5 Two-dimensional graph of the strength of the H_1 field in a 300 MHz probe as a function of the vertical position (z) in the coil. The method used is described in the text.

1.5. Different kinds of modulation

As has been pointed out earlier, the principle of two-dimensional spectroscopy relies on the fact that spins detected during a time t_2 remember what happened to them during a previous time interval with length t_1. This history can effect either the phase or the amplitude of the magnetization corresponding to these spins which is detected during the time t_2. These two kinds of modulation are known as phase modulation and amplitude modulation respectively (11). In some experiments a combination of these two kinds of modulation arises. In the case of phase modulation it is possible to determine the sign of modulation frequency, while in the case of amplitude modulation this is generally not possible. Determination of the sign of the modulation frequency is important in almost all 2D experiments since usually some are positive and some are negative if the transmitter frequency is placed in the centre of the spectral region of interest (see e.g. sections 2.2.3 and 2.3.2). Furthermore, distinguishing between the different kinds of modulation is of importance, because it determines the kind of line shape which can eventually be obtained in the 2D spectrum.

1.5.1 Phase modulation

In the case of phase modulation (see e.g. section 3.1) the phase of a magnetization component at the beginning of the detection period is a linear function of the length t_1 of the evolution period. The time domain signal can then be written as

$$s(t_1,t_2) = C \exp(i\Omega_1 t_1)\exp(i\Omega_2 t_2)\exp(-t_1/T_2^{(1)})\exp(-t_2/T_2^{(2)}) \qquad [1.45]$$

in which C is a complex amplitude factor, and $\exp(i\Omega_1 t_1)$ expresses the phase dependence on t_1 of the signal which is detected during t_2 and rotates with angular frequency Ω_2. The decay of the signal during evolution and detection period is expressed by the decay constants $T_2^{(1)}$ and $T_2^{(2)}$ respectively.

A complex two-dimensional Fourier transformation can be considered as two successive one-dimensional final time Fourier transformations:

$$S(\omega_1,\omega_2) = C \int_0^{FT_1} \exp(i\omega_1 t_1) \int_0^{FT_2} s(t_1,t_2)\exp(i\omega_2 t_2)dt_2 dt_1, \qquad [1.46]$$

where FT_1 and FT_2 are the Fourier transformation times along the t_1-and t_2-axis.

Fourier transformation with respect to t_2 gives:

$$S(t_1,\omega_2) = C \exp(i\Omega_1 t_1)\{A_2(\omega_2) + iD_2(\omega_2)\}\exp(-t_1/T_2^{(1)}) \qquad [1.47]$$

with $A_2(\omega_2)$ and $D_2(\omega_2)$ defined in Eq.[1.40].

This represents a set of spectra for different values of t_1, with phase $\Omega_1 t_1$. Fig.1.6 shows the real part of the Fourier transform for the case of a positive and a negative value of Ω_1. The negative frequency corresponds to a right hand screw, a positive Ω_1 corresponds to a left hand screw. Information about the sign of the modulation frequency is thus present.

A complex Fourier transformation of Eq.[1.47] with respect to t_1 gives

$$S(\omega_1,\omega_2) = C\{A_1(\omega_1)A_2(\omega_2) - D_1(\omega_1)D_2(\omega_2) + iA_1(\omega_1)D_2(\omega_2) + \\ + iD_1(\omega_1)A_2(\omega_2)\}. \qquad [1.48]$$

As can be seen from this equation, the sign of the modulation frequency is automatically determined. The practical realization of this complex two-dimensional Fourier transformation is explained in section 1.8. Supposing that the instrumental constant C is adjusted to be real, the real part of Eq. [1.48] represents a complicated line shape which consists of a superposition of a two-dimensional absorption and a two-dimensional dispersion

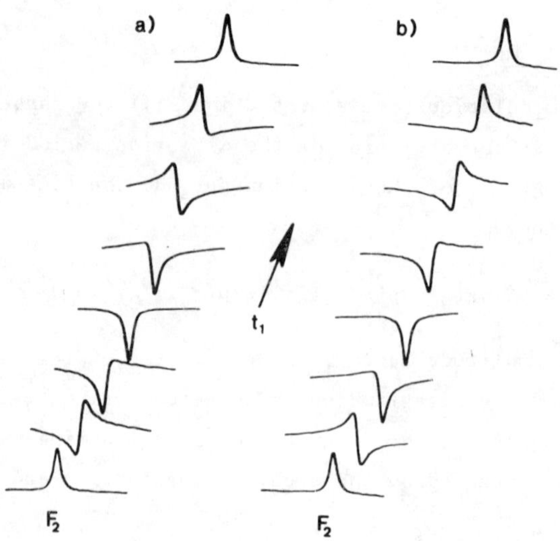

Fig.1.6 A resonance line modulated in phase as a function of t_1 by (a) a negative and (b) a positive modulation frequency.

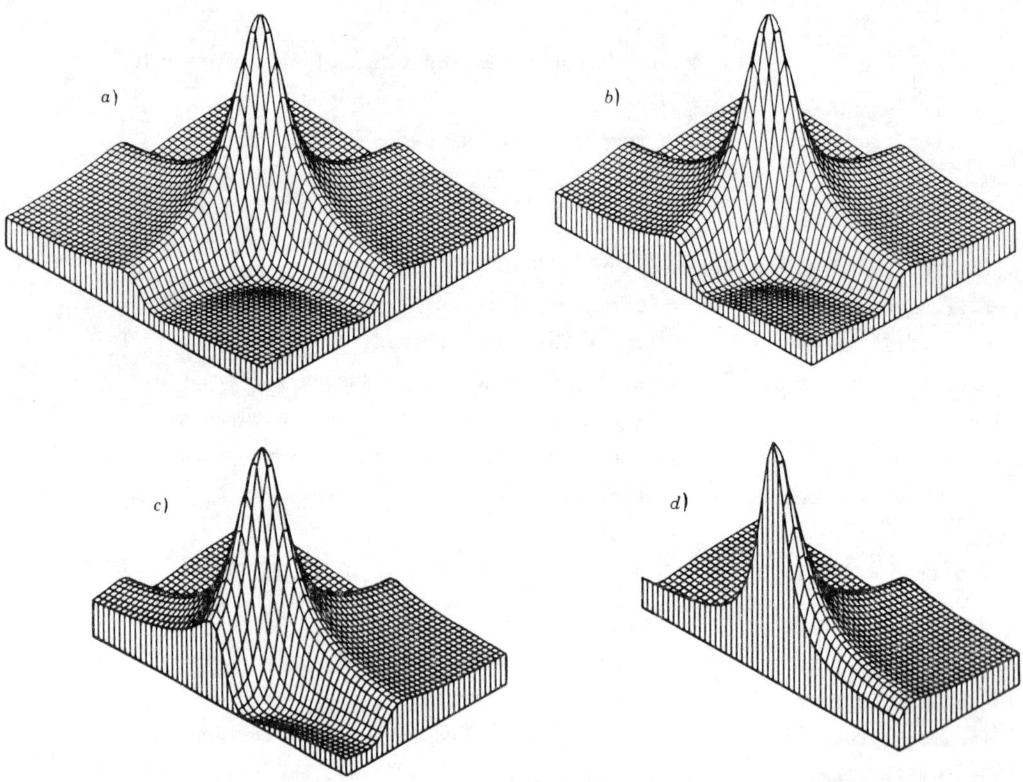

Fig.1.7 Computer simulations of the phase twist line shape obtained in many two-dimensional NMR experiments. A vertical section through this solid shows a dispersion mode shape, far from resonance, (a), with a gradual progression (b,c) towards a pure absorption mode section at exact resonance, (d). From ref.(11).

line shape (Fig.1.7).

It appears to be impossible in general to modify this so-called phase-twisted line shape into pure two-dimensional absorption by means of any simple phase correction routine. The width of this line is thus essentially determined by the width of the broad dispersion contribution.

If besides the signal of Eq.[1.45] another so-called reversed precession signal of the form

$$s(t_1,t_2) = C \exp(-i\Omega_1 t_1)\exp(i\Omega_2 t_2)\exp(-t_1/T_2^{(1)})\exp(-t_2/T_2^{(2)}) \qquad [1.49]$$

is also available, adding or subtracting the two signals causes cancellation of the dispersion contribution (12). In fact, if the addition or subtraction is done before Fourier transformation, this becomes a case of amplitude modulation. It is a general result that an amplitude-modulated signal can be decomposed into the sum of two phase-modulated signals★.

1.5.2 Amplitude modulation

If the amplitude of a magnetization component in the transverse plane at the beginning of the detection period is an oscillating function of the time t_1, while its phase is independent of t_1, this is called amplitude modulation (see e.g. section 1.4). The signal is then described by

$$s(t_1,t_2) = C \cos(\Omega_1 t_1)\exp(i\Omega_2 t_2)\exp(-t_1/T_2^{(1)})\exp(-t_2/T_2^{(2)}) . \qquad [1.50]$$

Again the two-dimensional Fourier transformation is split into two one-dimensional Fourier transformations. After Fourier transformation with respect to t_2, and provided that the instrumental constant C is real, a set of pure absorption and dispersion spectra is obtained:

$$S(t_1,\omega_2) = C \cos(\Omega_1 t_1)\{A_2(\omega_2) + iD_2(\omega_2)\}\exp(-t_1/T_2^{(1)})\} . \qquad [1.51]$$

An example of these amplitude modulated absorption spectra is shown in Fig.1.4a.

A real Fourier transformation of the real part of Eq.[1.51] with respect to t_1 gives

$$S(\omega_1,\omega_2) = C A_2(\omega_2)\{A_1(\omega_1) + A_1(-\omega_1) + iD_1(\omega_1) + iD_1(-\omega_1)\}. \qquad [1.52]$$

★ This follows immediately from the mathematical relation: $\exp(-i\Omega_1 t_1) + \exp(i\Omega_1 t_1) = 2\cos(\Omega_1 t_1)$.

As is clear from this equation it is impossible to determine whether Ω_1 was positive or negative. This agrees with the earlier statement that an amplitude-modulated signal can always be considered as the sum of two signals modulated in phase by opposite frequencies.

In the case of amplitude modulation it is thus advantageous to use a real instead of a complex Fourier transformation with respect to t_1 after phasing the real part of the ω_2-spectra to the absorption mode, since this gives two-dimensional absorption line shapes.

1.5.3 Phase- and amplitude modulation

In some experiments both the amplitude and the phase of a detected magnetization are modulated as a function of t_1 (see e.g. section 3.2.3). In this case the time domain signal is written as

$$s(t_1,t_2) = C\cos(\Omega_{1A}t_1)\exp(i\Omega_{1B}t_1)\exp(i\Omega_2 t_2), \qquad [1.53]$$

where Ω_{1A} and Ω_{1B} are the frequencies with which the signal is modulated in amplitude and in phase.

However, this signal can always be considered as the sum of two components:

$$s(t_1,t_2) = \tfrac{1}{2}C\exp\{i(\Omega_{1B}-\Omega_{1A})t_1\}\exp(i\Omega_2 t_2) + \tfrac{1}{2}C\exp\{i(\Omega_{1B}+\Omega_{1A})t_1\}\exp(i\Omega_2 t_2). \qquad [1.54]$$

Both components are modulated in phase as a function of t_1, and a complex two-dimensional Fourier transformation will generate two phase-twisted resonance lines at $(\Omega_{1B}-\Omega_{1A}, \Omega_2)$ and $(\Omega_{1B}+\Omega_{1A}, \Omega_2)$.

1.6 Presentation of two-dimensional spectra

With the development of two-dimensional spectroscopy a new problem arises, because an intensity now has to be presented as a function of two independent frequency parameters. One problem is the complicated phase-twisted line shape as described in the previous section, and another problem is the graphical presentation which is more complicated than in the one-dimensional case.

1.6.1 Physical presentation

As mentioned before a 2D Fourier transformation can be considered as consisting of two one-dimensional Fourier transformations, usually first with respect to t_2 and then with respect to t_1. After the first Fourier transformation a set of complex data will always be left. One can use the complex data as the input for the second, complex Fourier transformation, or one can use the real or imaginary part as the input for the second real Fourier transformation with respect to t_1. In both cases this gives a set of complex data points:

$$S(\omega_1,\omega_2) = R(\omega_1,\omega_2) + iI(\omega_1,\omega_2), \qquad [1.55]$$

where both R and I can contain a mixture of absorptive and dispersive signals.

Phase-sensitive mode

If a pure two-dimensional absorption spectrum $R(\omega_1,\omega_2)$ can be obtained (section 1.5.2) it is nearly always advantageous to plot this absorption spectrum in the phase-sensitive mode. The advantages compared with the other modes are the high resolution of absorption peaks (only compared with the absolute value mode), the possibility of distinguishing between positive and negative intensities, and a better sensitivity. If only one cross-section through the two-dimensional spectrum is displayed (section 1.6.2) it is nearly always possible to adjust the phase of the resonances to absorption or to an absorption-like mode, while the rest of the spectrum remains unphased. This is even possible in the case of phase modulation, where it is impossible in general to obtain a two-dimensional absorption display of the whole spectrum. In the latter case it can be advantageous to present the whole spectrum in the power or in the absolute value mode.

Power mode

The power mode spectrum $S^2(\omega_1,\omega_2)$ is defined as

$$S^2(\omega_1,\omega_2) = R^2(\omega_1,\omega_2) + I^2(\omega_1,\omega_2). \qquad [1.56]$$

The line shape of a single Lorentzian line in the power mode is identical to the line shape of the corresponding absorption mode, so high resolution is still obtained. A basic problem is however, that while overlap of absorption mode components is additive, overlap of dispersion mode is subtractive, so that quite gross line shape distortions can occur in the case of overlapping resonances (13). Another disadvantage is that the normal intensities are squared. This means that very large intensity distortions can occur, which especially in two-dimensional spectroscopy can be rather inconvenient. Therefore, absolute value mode presentation is usually preferred.

Absolute value mode

The absolute value mode is defined as the square root of the power mode:

$$|S|(\omega_1,\omega_2) = \{R^2(\omega_1,\omega_2) + I^2(\omega_1,\omega_2)\}^{\frac{1}{2}} . \qquad [1.57]$$

The intensity distortions of the power mode are obviously not present in this presentation; however, the line shape distortion as found in the power mode is still present. The line shapes are much less attractive because of their large width, especially at the base of the line. For Lorentzian lines the broadening of the absolute value mode compared with the absorption mode is a factor √3 at half height and a factor 10 at 1% height.

Nevertheless, the absolute value mode is the most commonly used mode to present 2D spectra to date. As explained in section 1.7.3 the use of digital filtering can remove the tailing of the absolute value line shape.

1.6.2 Graphical presentation

There are several ways to represent a spectrum as a function of two independent variables. The most common methods are the so-called stacked-trace plot (e.g.Fig.1.5) and the contour plot (14) (e.g. Fig.1.11). In principle a coloured plot or a dark-light intensity plot as used in spin-imaging (15,16) can be used as well, but these are not commonly used at present. Cross-sections and projections (17) are very useful for extracting quantitative information from a 2D spectrum.

Stacked-trace plot

Many spectra are drawn behind each other as a function of one of the frequency variables, for a set of incrementing values of the other frequency, in such a way that a three-dimensional looking graph appears, where the third dimension indicates the intensity of the function. To give a better three-dimensional impression, a so-called white-wash procedure is usually applied. This procedure ensures that parts of traces which have an intensity so low that they would be below parts of previous traces are not plotted. Fig.1.5 and 1.11 are examples of stacked-trace plots without and with white-wash procedure. Despite the attractive appearance of stacked-trace spectra, it is in practice hard to extract information since frequency coordinates of peaks are hard to determine. Another disadvantage is that small peaks can be hidden completely behind large ones. Making a stacked-trace plot is often rather time-consuming and can take up to a few hours, depending on the size of the data matrix and the kind of hard- and software available.

Contour plot

In contour plots only the contours of a peak are plotted, similar to altitude lines on a geographical map. If the lowest contour is defined to appear at a certain signal amplitude (altitude) C, the next contours are usually plotted for intensity 2C, 4C, 8C, etc. This allows a high dynamic range without plotting too large a number of contours. It is often possible to extract quantitative information about frequencies from such a contour plot, while at the same time an impression of the amplitude is obtained by counting the number of contours around a certain peak. A problem with contour plots is that the lowest contours of intense resonances which have the absolute value line shape, extend over a large area in the two-dimensional spectrum, obscuring the presence of resonances with low intensity, or giving rise to small spurious peaks due to interference between two resonances. Making a contour plot generally takes considerably less time than a stacked-trace plot. In Fig. 1.11 an example of a contour plot can be found.

Cross-sections

In practice the peaks with the crucial information lie on a limited number of cross-sections in the 2D spectrum. Certainly, if a detailed look at these peaks is required, and quantitative results like frequency or amplitude have to be extracted, it is advisable to plot these different traces as separate spectra (17). The cross-sections required are mostly parallel to one of the main frequency axes, or as happens in homonuclear J-resolved spectroscopy (section 3.3), make an angle of 45° with these axes.

Projections

In homonuclear J-resolved spectroscopy, multiple quantum spectroscopy (chapter 4) and in homonuclear shift correlation spectroscopy (section 2.3.6) it can be advantageous to calculate a projection of the 2D spectrum by integrating the intensity of cross-sections perpendicular to the axis on which the projection takes place.

The main advantages and disadvantages of the different display modes are summarized in Table 1.1. In practical cases often more than one display mode of a certain spectrum is used to extract most information.

Table 1.1 Advantages and disadvantages of four different display modes.

	Advantages	Disadvantages
Stacked-trace	plots all information in a suggestive way.	hidden peaks; time-consuming; hard to extract frequencies.
Contour	easy to extract frequencies; relatively fast.	information below lowest contour and in between contours gets lost.
Cross-section	easy to extract frequencies, amplitudes and line shapes.	only part of the information.
Projection	simplification; easy to extract frequencies and line shapes.	only part of the information.

1.7 The effects of digital filtering

In this paragraph three different digital filtering techniques, already commonly used in one-dimensional NMR spectroscopy, will be discussed. These three filters are: exponential weighting, Lorentzian to Gaussian transformation (product of an exponential and a Gaussian weighting), and the convolution difference filter (difference between no weighting and an exponential weighting). The effect of these filters on sensitivity and resolution in one- and two-dimensional spectroscopy will be considered. No attention will be paid to the influence of the experiment repetition rate on the sensitivity.

In a paper by Aue et al. (18) it has been shown that the signal-to-noise ratio in two-dimensional spectroscopy can be of the same order of magnitude as in one-dimensional spectroscopy, provided that the following conditions are met:

- matched digital filtering (see e.g. ref.19),
- length of acquisition and delay times optimized for sensitivity,
- equal numbers of peaks in the one- and two-dimensional spectra.

For various reasons, in many practical applications of the two-dimensional methods these conditions cannot be satisfied, and signal-to-noise ratio will be significantly less. It is for instance often not possible to apply matched filtering because of the corresponding line-broadening effect. Quite frequently the opposite - strong resolution enhancement - is required along one or both axes in the time domain data matrix. This is needed to show features which otherwise would be hidden because of overlap of resonances in the two-dimensional spectrum. Since many real applications of two-dimensional spectroscopy are connected with solving problems concerning rather complicated molecules with an extensive overlap between the different resonances, the question of digital filtering is often of the utmost importance. It will appear that in order to obtain a certain resolution enhancement in both dimensions of the 2D spectrum, as is often necessary, a much larger sensitivity loss has to be accepted compared with the same narrowing in one-dimensional spectroscopy.

The effect of a one-dimensional filtering function g(t) is to multiply the time domain signal at time t by the corresponding function value g(t). Similarly, a two-dimensional filtering function $g(t_1,t_2)$ can be defined. All the filtering functions described in this paragraph $g(t_1,t_2)$ can be considered as the product of two independent one-dimensional weighting functions:

$$g(t_1,t_2) = g_1(t_1) \times g_2(t_2) \ . \qquad [1.58]$$

This means that the actual two-dimensional weighting is done in two steps: First the responses acquired for the different values of t_1 are all multiplied by the same weighting function $g_2(t_2)$, second - after Fourier transformation and transposition - all the new rows of the data matrix containing the modulation information are multiplied by weighting function $g_1(t_1)$. Using this principle the effect of the filtering function $g(t_1,t_2)$ on the signals and noise in two-dimensional spectroscopy can easily be understood, and only a discussion of the one-dimensional filtering functions is needed.

If the noise in the acquired responses is stationary and uncorrelated the filtering function $g_2(t_2)$ changes the average power P_N of the noise according to

$$P'_N = P_N/AT_2 \int_0^{AT_2} g_2(t_2)^2 dt_2, \qquad [1.59]$$

where AT_2 is the acquisition time along the t_2-axis★. In the mean-time the amplitude S of a resonance line, obtained after Fourier transformation of a time domain signal $s_2(t_2)$ changes its amplitude according to

★ If only part of the length of the acquired response is used in the processing, the time $AT_{1,2}$ should be adjusted correspondingly.

$$S' = \frac{S \int_0^{AT_2} |s_2|(t_2)g_2(t_2)dt_2}{\int_0^{AT_2} |s_2|(t_2)dt_2} \qquad [1.60]$$

The signal-to-noise ratio in the frequency domain will be proportional to $S/\surd P'_N$. Since after this weighting and Fourier transformation with respect to t_2 the noise along the t_1-axis is still uncorrelated and stationary, the effect of the filtering function $g_1(t_1)$ on the signal amplitude and noise power can be calculated in a similar way:

$$P''_N = P'_N/AT_1 \int_0^{AT_1} g_1(t_1)^2 dt_1 \; . \qquad [1.61]$$

$$S'' = \frac{S' \int_0^{AT_1} |s_1|(t_1)g_1(t_1)dt_1}{\int_0^{AT_1} |s_1|(t_1)dt_1} \; . \qquad [1.62]$$

The eventual signal-to-noise ratio in the 2D spectrum will be proportional to $S''/\surd P''_N$. It then follows from Eqs.[1.59] to [1.62] that a change in sensitivity by a factor C_2 after the first weighting and Fourier transformation, followed by a change C_1 due to the second weighting and Fourier transformation, gives a change of sensitivity in the 2D spectrum equal to $C_1 \times C_2$.

In the following discussions of the individual filters, the effects on an exponentially decaying signal with time constant T_2 and angular frequency Ω will be treated:

$$s(t) = s(0)\exp(i\Omega t)\exp(-t/T_2) \; . \qquad [1.63]$$

Note however, that the approximation of exponential decay is in practice often not strictly valid and changes in the results will occur.

An important consideration using digital filtering is that the signal amplitude should have decayed sufficiently {in practice by a factor $20 \approx \exp(3)$} at the end of the time domain signal, after multiplying with the filter function, in order to avoid truncation and corresponding artefacts in the spectrum. This principle will be used in all quantitative analyses given below. As a consequence of this, it is generally necessary in the case of line-narrowing along the t_1-axis to employ longer acquisition times along the t_1-axis, even though the responses for long t_1-values have bad sensitivity. The prolonging of the experiment causes an extra loss in sensitivity which will not be taken into account in the discussions given below.

1.7.1 Exponential weighting

Multiplication of the signal of Eq.[1.63] with the filtering function

$$g(t) = \exp(t/T_L) \quad [1.64]$$

yields again an exponentially decaying signal:

$$s'(t) = s(0)\exp(i\Omega t)\exp(-t/T_2') \quad [1.65]$$

with

$$(T_2')^{-1} = (T_2)^{-1} - (T_L)^{-1}.$$

The new decay constant T_2' determines the line width in the frequency domain. If the constant T_L is positive the signal will decay slowlier and the resolution will be enhanced. A negative value for T_L will increase the decay rate and decrease the resolution but increase sensitivity. Optimum sensitivity is obtained if $T_L = -T_2$ (matched weighting). Neglecting the truncation at the end of the time domain signal, the frequency spectrum obtained after Fourier transformation of the signal from Eq.[1.65] is given by

$$S(\omega) = s(0)\left[\frac{T_2'}{1 + T_2'^2(\Omega - \omega)^2} + \frac{i\,T_2'^2(\Omega - \omega)}{1 + T_2'^2(\Omega - \omega)^2}\right]. \quad [1.66]$$

For the real part of this spectrum this implies a width at half height given by

$$\Delta\nu_{0.5} = 1/\pi T_2' \ . \quad [1.67a]$$

At 1% of the peak height the width equals

$$\Delta\nu_{0.01} = 9.95/\pi T_2' \ . \quad [1.67b]$$

The use of an increasing exponential filter for resolution enhancement is not simple because of the disastrous effects it can have on the sensitivity if improperly used (20), and because of the artefacts arising in the case of truncation.

The strategy recommended in this study for using the exponential filter for line narrowing, is the following:

(a) choose a resolution of r Hz which one wants to get eventually, where r is related to T_2' by the relation $r = 1/\pi T_2'$.

(b) use only the first $3/\pi r \approx r^{-1}$ seconds of the time domain signal for the

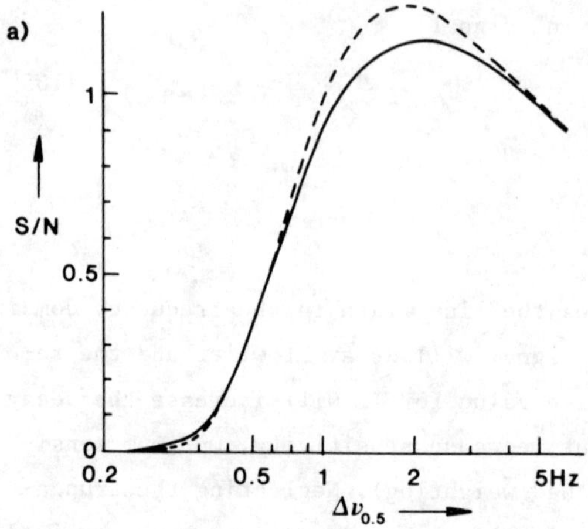

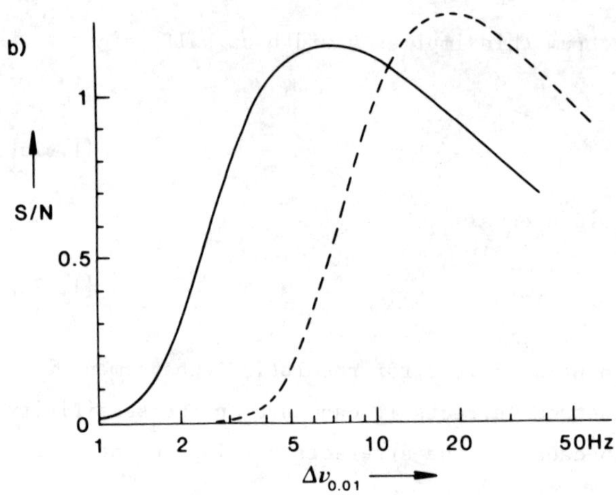

Fig.1.8 The relation between the signal-to-noise ratio and the line width at half height (a) and at 1% of the maximum height (b) if exponential filtering (broken line) or Lorentzian to Gaussian transformation (drawn line) is applied to a signal that corresponds to a Lorentz line with $\Delta\nu_{0.5}$ = 1Hz.

weighting and Fourier transformation. (Of course, an optimum choice for the length of the acquisition time would have been r^{-1} sec.) Use up to ten times zero filling (20) before the Fourier transformation, if sufficient data storage space is available, for proper digitization of the frequency domain.

(c) If the value of T_2 is known, T_L can be directly determined from:

$$T_L = T_2/(1 - r\pi T_2). \qquad [1.68]$$

If T_2 is not known, a minimum value for the time constant T_L greater than T_2 has to be selected which still gives acceptable truncation effects in the frequency domain spectrum.

(d) If sensitivity turns out to be too poor, repeat the process for a larger value of r.

Using this procedure, the effect on sensitivity for a certain factor of line narrowing can be found in the broken lines in Fig.1.8. The optimal use of the exponential filter becomes very difficult if resonances with different widths are present in the same spectrum, as is commonly the case.

The two-dimensional Lorentzian line shape shows in both the absorption and absolute value mode a pronounced star shape (section 1.7.5).

1.7.2 Lorentzian to Gaussian transformation

The exponential decay of a signal can be converted into a Gaussian decay, by multiplication of the signal with the filtering function

$$g(t) = \exp(t/T_L)\exp(-t^2/T_G^2) \qquad [1.69]$$

and setting $T_L = T_2$. Eq.[1.69] is the general form for the Lorentzian to Gaussian transformation filter as proposed for use in NMR by Lindon and Ferrige (21,22). If T_L is chosen to be equal to T_2 the time domain signal becomes

$$s'(t) = s(0)\exp(i\Omega t)\exp(-t^2/T_G^2) \qquad [1.70]$$

which after Fourier transformation gives for the real part

$$S(\omega) = \tfrac{1}{2}\sqrt{\pi}\, s(0)T_G\exp\{-(\Omega - \omega)^2T_G^2/4\}. \qquad [1.71]$$

This implies a line width at half height

$$\Delta\nu_{0.5} = 1.67/\pi T_G \qquad [1.72a]$$

and at 1% of the maximum height

$$\Delta\nu_{0.01} = 3.46/\pi T_G \; . \qquad [1.72b]$$

For the imaginary part of the Fourier transform no simple analytical expression is available. Obviously, this filter can be used for line narrowing by setting T_G to a large enough value. As is visible in Fig.1.8 the decrease in sensitivity for a certain amount of line narrowing at half height is of the same order as for exponential narrowing. Note however, that at 1% height the Gaussian line is 4.8 times narrower than the Lorentzian line, if both resonances have identical width at half height.

The Lorentzian to Gaussian transformation filter is easier to use than exponential narrowing because the length of the time domain signal does not really matter for the purpose of sensitivity, provided that it is longer than $\sqrt{3}\, T_G$, since the extra length can be considered as zeroes, due to the very small value of g(t) at the end of the time domain signal. Zero filling of the time domain signal until it corresponds to a length of $15T_G$ can be needed for proper digitization of the frequency spectrum. If the time constant T_L in Eq.[1.69] is chosen slightly shorter than the decay constant T_2 of the signal, a dip in the base line of the spectrum at the position of the resonance will occur. This effect can be used in an iterative process to make a proper choice for T_L. Of course, problems arise again in using the filter if resonances in the same spectrum have different width. However, because no truncation occurs, the interference with other narrower lines which are over-enhanced, is usually not much of a problem.

Apart from line narrowing the filter has three other important applications:
(1) In the case where the sampling time along one or both time axes is limited to shorter than about $1.5T_2$, e.g. because of limited data storage space, Gaussian instead of exponential weighting should be used to avoid truncation, yielding better sensitivity and resolution.
(2) In the case of unresolved multiplets, the decay of the time domain signal is closer to Gaussian than exponential and hence, Gaussian weighting can be closer to matched filtering, yielding better sensitivity. Of course, this is only true if no attempt is made to resolve the multiplet.
(3) The filter can be used for the transformation of an exponentially decaying signal into a pseudo-echo as explained in the next section.

The two-dimensional absorption Gaussian line shape has circular or ellipsoidal contours, but the absolute value mode presentation shows a pronounced star shape with strong tailing, as shown in section 1.7.5.

1.7.3 The pseudo-echo transformation

Many 2D spectra are usually presented in the abolute value mode, either to avoid a sometimes cumbersome phasing of the two-dimensional spectrum, or just because for many methods no way is known to obtain a pure absorption mode spectrum. Absolute value lines usually have an unattractive line shape with wide wings. It will be shown that if the signal is filtered in a special way, which gives the envelope a shape which is symmetrical in time about the midpoint of the time domain signal, no dispersive components are obtained after Fourier transformation (23,24) which results in a higher resolution in the absolute value mode.

Consider a signal of the form

$$s(t) = \exp(i\Omega t + i\phi)f(t) \ , \qquad [1.73]$$

where f(t) describes the decay of the signal, and f(t) is symmetric about time zero.

This can be rewritten as

$$s(t) = \exp(i\phi)\exp(i\Omega t)f(t). \qquad [1.74]$$

Final time Fourier transformation of this signal gives

$$S(\omega) = \int_{-T}^{T} s(t)\exp(i\omega t)dt \qquad [1.75]$$

$$= \exp(i\phi)\int_{-T}^{T} \cos(\Omega t)\cos(\omega t)f(t)dt + i\exp(i\phi)\int_{-T}^{T} \sin(\Omega t)\sin(\omega t)f(t)dt$$

since the integral over the product of an odd and an even function is automatically zero, e.g.

$$\int_{-T}^{T} \cos(\Omega t)\sin(\omega t)f(t)dt = 0. \qquad [1.76]$$

Since both products in the right half of Eq.[1.75] give absorptive contributions only, with amplitudes proportional to $\exp(i\phi)$, an absolute value calculation will give an absorptive line shape with amplitude independent of ϕ. The line shape will depend on the decay function f(t). The method relies on the symmetry of f(t) around time zero. A function f(t) symmetrically decaying about a certain time T_c gives similar results; this case can be considered as a shift of the origin of the time domain, inducing only a change in phase which disappears after the absolute value calculation. A condition for absorptive absolute value line shapes is that the signal amplitude at the beginning and the end of the signal is sufficiently close to zero. If the signal is truncated at one end only, the absolute value line shape again has partly absolute value character, depending on the severity of the truncation. If the signal is truncated at both ends, base line distortions (sinc-functions) will appear in the frequency spectrum.

To obtain a symmetrical envelope amplitude, the time domain signal (Fig. 1.9a) is in practice first multiplied with an increasing exponential to eliminate the decay (Fig.1.9b). Second, the modified response is weighted by a decreasing exponential or Gaussian function, symmetrical about its midpoint T_c to yield a so-called pseudo-echo shape (1.9c). The overall filter can then be written as

$$g(t) = \exp(t/T_2)\exp(-|t - T_c|/T_L) \quad [1.77a]$$

or

$$g(t) = \exp(t/T_2)\exp\{-(t - T_c)^2/T_G^2)\}. \quad [1.77b]$$

The latter pseudo Gaussian echo filter is usually preferred because of the narrow base and the attractive circular contours of absolute value line shapes. The time constant T_G is chosen to be smaller than or equal to $T_c/\sqrt{3}$ to avoid truncation. It can be shown by substitution that in the case where

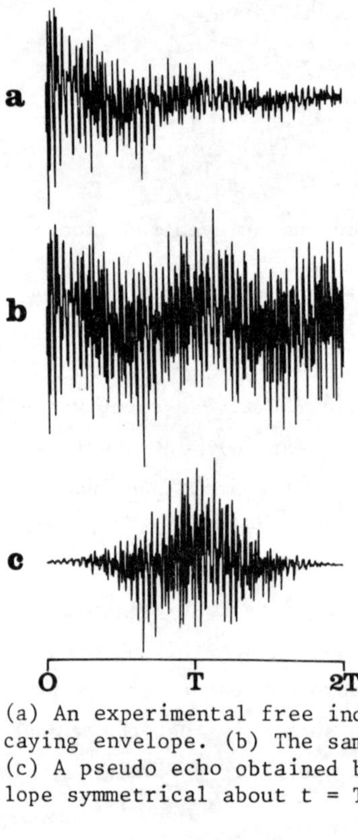

Fig.1.9 (a) An experimental free induction signal with an exponential decaying envelope. (b) The same signal multiplied by $\exp(t/T_2)$. (c) A pseudo echo obtained by reshaping (b) to give a Gaussian envelope symmetrical about t = T.

$T_G = T_c/\sqrt{3}$ the window of Eq.[1.77b] gives the same results as the Lorentzian to Gaussian transformation filter (Eq.[1.69]) with time constants $(T_L)^{-1} = (T_2)^{-1} + (T_c/6)^{-1}$ and identical value for T_G.

The pseudo-echo window should be used with care, since large intensity anomalies can occur in the case of different relaxation times and in the case where lines are not resolved, i.e. if the line separation is less than $(T_c)^{-1}$ for the pseudo Gaussian echo filter.

The loss in sensitivity due to the use of the pseudo-echo filter will usually be very large although in certain cases no loss or even a gain in sensitivity occurs (section 2.3.2).

The pseudo-echo filter should only be used in the case where an absolute value mode display is needed.

1.7.4 The convolution difference filter

The convolution difference filter is described as

$$g(t) = 1 - \exp(-t/T_{CD}) \ . \qquad [1.78]$$

It was first proposed by Campbell et al. (25), and is generally used to eliminate broad background signals as often encountered in biological systems and liquid crystals. The idea is to apply strong exponential weighting with time constant T_{CD} which leaves only the broad background signals. This signal is then subtracted from the original signal, leaving only the slowly decaying components. So the convolution difference filter is thus not normally used for line narrowing, but for elimination of unwanted broad components. If the time constant T_{CD} is chosen to be short compared with the decay constant of the signal, the loss in sensitivity will be quite small. After applying the convolution difference filter the filtered signal can be filtered once more by a decaying exponential or Gaussian function to increase the sensitivity.

The magnitude of the dispersive component at large offsets of the centre of a Lorentzian resonance line does not depend on the decay constant T_2 of the time domain signal (Eq.[1.40]). By using the convolution difference filter, which gives the difference of two Lorentzian resonance lines, the dispersive component is thus eliminated at large offsets from the centre of the line. Therefore, in the absolute value mode, tailing will be partially suppressed. The effectiveness of the suppression of tailing by using the convolution difference filter depends on the time constant T_{CD} in Eq.[1.78]: the longer T_{CD}, the more effective the suppression of tailing but the larger the loss in sensitivity.

In two-dimensional spectroscopy the main advantages in using the filter are:

(1) The absolute value tailing is partially suppressed.

(2) The filter can be used advantageously in homonuclear correlation experiments to decrease the intensity of diagonal peaks in the 2D spectrum and at the same time emphasize the presence of cross-peaks (7) as explained in section 2.3.2.

1.7.5 Comparison of line shapes

To illustrate the effect of using different filters on the line shape, in Fig.1.10 the different one-dimensional absorption mode line shapes and the corresponding absolute value mode are displayed. Filtering constants have

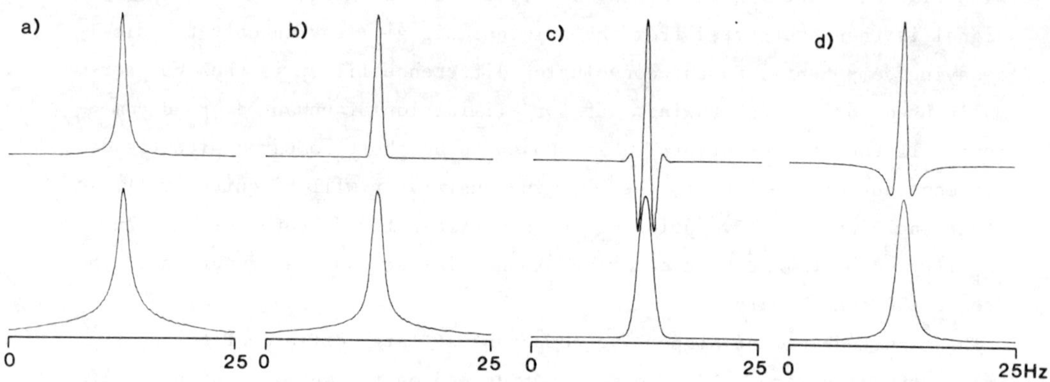

Fig.1.10 Computer simulations of different line shapes, top the absorption mode and bottom the corresponding absolute value mode, for different decay functions of the time domain signal. (a) A Lorentzian decay, (b) a Gaussian decay, (c) a Gaussian pseudo echo and (d) a convolution difference signal.

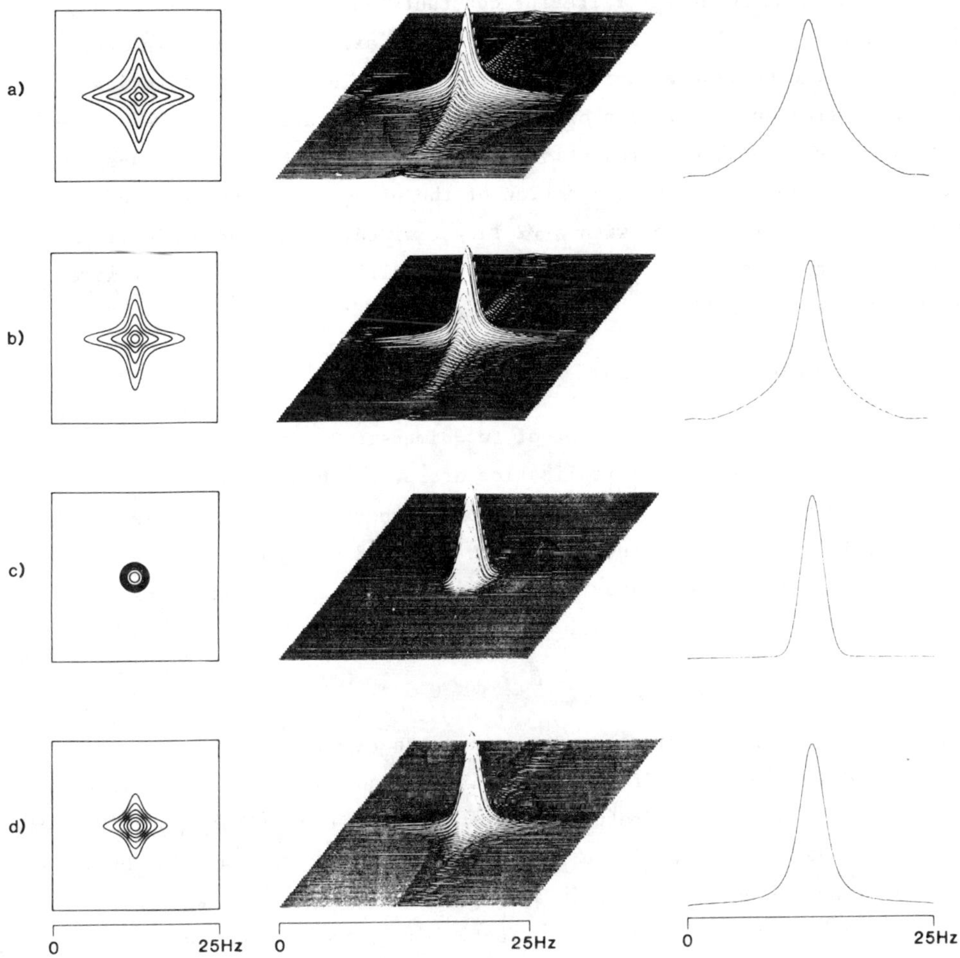

Fig.1.11 Computer simulations of different absolute value mode line shapes. At the left hand side the contour plot of the two-dimensional line shape is shown, in the middle a stacked trace plot is shown and at the right hand side the projection of the line along the diagonal onto the horizontal axis is presented.
The line shapes have been obtained by using
(a) an exponential filter,
(b) Lorentzian to Gaussian transformation,
(c) a pseudo Gaussian echo transformation,
(d) a convolution difference filter.

been chosen in such a way that all signals have decayed to 1/20 of their maximum value in one second.

Fig.1.11 shows the absolute value mode display of the two-dimensional line shapes using identical filtering constants as in Fig.1.10. In this figure also the results of a projection along an axis which makes an angle of 45° with the horizontal axis are shown. Each point of the projection is calculated from an integration of the cross-section parallel to the direction of projection. These projections are important in homonuclear J-spectroscopy (section 3.3). The projection of the pseudo Gaussian resonance is another Gaussian line shape with a $\sqrt{2}$ larger width. All other projections result in very complicated expressions for the line shapes which all have extremely wide wings (17).

1.8 Experimental realization

With the theoretical development of two-dimensional spectroscopy new problems about the practical realization arose. In this section the operations necessary to perform a two-dimensional experiment will be briefly analysed. In order to show the extensions required compared with the conventional one-dimensional experiment, a similar survey of operations for one-dimensional spectroscopy will be given first.

One-dimensional spectroscopy

In one-dimensional Fourier transformation experiments, a pulse response is detected as a function of a running time variable t. Either real (single detection) or complex (quadrature detection) sampling points are acquired at times $n\Delta t$, where n is a positive integer and $(\Delta t)^{-1}$ is the sampling frequency. This digitized signal is stored in a computer memory or on background storage. In order to improve sensitivity by a factor $\sqrt{p}$, the experiment can be repeated p times, coadding the results. Digital filtering of the acquired response can be performed to increase the resolution or sensitivity in the eventual frequency spectrum. Digital Fourier transformation, either real or complex, depending on whether single or quadrature detection was used, yields the digitized frequency spectrum. The number of data points prior to Fourier transformation may be increased up to a factor 10 by adding zeroes to the end of the response, in order to obtain a better discretization of the frequency spectrum (20).

Two-dimensional spectroscopy

In a two-dimensional experiment, the pulse sequence generating the response of interest is repeated for a large number of equally incremented values for the length t_1 of the evolution period. Each of the responses is acquired, either real or complex, during a time t_2, digitized and generally stored on the background storage (e.g. magnetic disc) of the computer. Time-averaging can be used again to improve sensitivity: either each sub-experiment with a certain value of t_1 is repeated a number of times p or the complete experiment - the total of all sub-experiments with different values for t_1 - is repeated p times. The latter way is preferred since slow variations (on the time scale of the entire time averaged experiment) partly average out. The increment in the length t_1 of the evolution period (Δt_1) determines the sampling frequency $(\Delta t_1)^{-1}$ along the t_1-axis.

The experiment thus produces a digitized data matrix $s(t_1,t_2)$ with sampling points at $(t_1,t_2) = (n\Delta t_1, m\Delta t_2)$, where n and m are positive integers.

A discrete Fourier transformation - real or complex, depending on whether single or quadrature detection during t_2 was used - will give a set of frequency spectra for the different values of t_1. These spectra can be considered as rows of a matrix. A column then contains information about the modulation as a function of t_1 (interferogram).

Since the first Fourier transformation with respect to t_2 produces complex data points, the interferograms consist of complex data. In order to have simple access to the interferograms, the matrix is usually transposed, leaving the complex interferograms as new rows of the data matrix. In the case of phase modulation a complex Fourier transformation of these interferograms is applied with respect to t_1, distinguishing between positive and negative modulation frequencies and giving phase-twisted line shapes.

In the case of amplitude modulation this complex Fourier transformation will also give proper results but the spectrum will be symmetrical about zero frequency in the F_1-dimension, and resonances will again have the phase-twisted line shape. To improve sensitivity and to produce a pure absorption mode two-dimensional spectrum, the real part of the F_2-spectra can be phased to the absorption mode before transposition. If the imaginary part of the data are then replaced by zeroes, the other parts of data-processing remain unchanged compared with the phase modulation case. A different approach in the case of amplitude modulation is to remove the dispersive imaginary part of the F_2-spectra after phasing and before transposition. A transposition of the real data will give interferograms consisting of real data. A real Fourier transformation with respect to t_1 will then produce the two-dimensional spectrum with only positive values for the F_1-dimension. This latter approach saves some processing time and data storage space.

Resolution along the F_2- or F_1-axis of the two-dimensional spectrum can be improved by digital filtering along the t_2- or t_1-axis before the respective Fourier transformations. It is noted here that resolution or sensitivity enhancement along the t_1-axis does not change the ratio between signal amplitude and the amplitude of so-called t_1-noise in the two-dimensional spectrum.

t_1-Noise is the extra noise in interferograms which contain signals, compared with interferograms which do not carry information. The t_1-noise arises from instabilities in the spectrometer system, e.g. changes in field inhomogeneity, field/frequency ratio, synthesizer noise and sample spinning.

References

(1) W.P. Aue, E. Bartholdi and R.R. Ernst, J. Chem. Phys. 64, 2229 (1976).

(2) C.P. Slichter, Principles of Magnetic Resonance, pp 127-134, Harper & Row, New York 1964.

(3) M. Goldman, Spin Temperature and Nuclear Magnetic Resonance in Solids, p 13, Clarendon Press, Oxford 1970.

(4) S. Schäublin, A. Höhener and R.R. Ernst, J. Magn. Reson. 13, 196 (1974).

(5) A.A. Maudsley, A. Wokaun and R.R. Ernst, Chem. Phys. Lett. 69, 567 (1979).

(6) K. Nagayama, A. Kumar, K. Wüthrich and R.R. Ernst, J. Magn. Reson. 40, 321 (1980).

(7) A. Bax, R. Freeman and G.A. Morris, J. Magn. Reson. 42, 162 (1981).

(8) A. Bax and G.A. Morris, J. Magn. Reson. 42, 501 (1981).

(9) A. Bax, P.G. de Jong, A.F. Mehlkopf and J. Smidt, Chem. Phys. Lett. 69, 567 (1980).

(10) F. Bloch, Phys. Rev. 70, 460 (1946).

(11) R. Freeman, Proc. R. Soc. Lond. A. 373, 149 (1980).

(12) P. Bachmann, W.P. Aue, L. Müller and R.R. Ernst, J. Magn. Reson. 28, 29 (1977).

(13) G. Bodenhausen and D.L. Turner, J. Magn. Reson. 41, 200 (1980).

(14) A. Kumar, D. Welti and R.R. Ernst, Naturwiss. 62, 34 (1975).

(15) W.S. Hinshaw, P.A. Bottomley and G.N. Holland, Nature 270, 722 (1977).

(16) P. Mansfield and A.A. Maudsley, Br. J. Radiology 50, 188 (1977).

(17) K. Nagayama, P. Bachmann. K. Wüthrich and R.R. Ernst, J. Magn. Reson. 31, 133 (1978).

(18) W.P. Aue, P. Bachmann, A. Wokaun and R.R. Ernst, J. Magn. Reson. 29, 523 (1978).

(19) D. Shaw, Fourier Transform NMR Spectroscopy, Elsevier, Amsterdam 1976.

(20) J.C. Lindon and A.G. Ferrige, Progr. in NMR Spectr. 14, 27 (1980).

(21) A.G. Ferrige and J.C. Lindon, J. Magn. Reson. 31, 337 (1978).

(22) J.C. Lindon and A.G. Ferrige, J. Magn. Reson. 36, 277 (1979).

(23) A. Bax, A.F. Mehlkopf and J. Smidt, J. Magn. Reson. 35, 373 (1979).

(24) A. Bax, R. Freeman and G.A. Morris, J. Magn. Reson. 43, 333 (1981).

(25) I.D. Campbell, C.M. Dobson, R.J.P. Williams and A.V. Xavier, J. Magn. Reson. 11, 172 (1973).

2 CHEMICAL SHIFT CORRELATION SPECTROSCOPY

2.1 Introduction

All two-dimensional shift correlation experiments have in common that they show a relation between different resonances, originating from an interaction between the corresponding nuclei. Although commonly used, the name shift correlation is essentially misleading, since not the shifts but the signals from nuclei with certain shifts are correlated. If in the two-dimensional spectrum a cross peak occurs at (Ω_A,Ω_B) this means that nucleus A and B have a mutual interaction. Nuclei A and B can be either of the same (homonuclear) or of different species (heteronuclear shift correlation). Depending on the type of experiment the kind of interaction shown is different. Possible mechanisms are homo- or heteronuclear scalar coupling (direct or long-range), chemical exchange, cross-relaxation and in non-isotropic samples dipole-dipole coupling. All correlation experiments, showing interaction between nuclei A and B, rely on partial magnetization transfer from nucleus A to B. However, the experimental realization varies so much for the different kinds of interaction that in the description which follows a subdivision will be made into three categories, chemical shift correlation through (a) heteronuclear scalar coupling, (b) homonuclear scalar coupling and (c) cross-relaxation and exchange. Chemical shift correlation through heteronuclear scalar coupling will be discussed in detail because of its practical importance, and because it can be described at an elementary level, using classical pictures. Relaxation and magnetic field inhomogeneity will be neglected throughout this chapter unless explicitly stated.

2.2 Correlation of chemical shifts through heteronuclear scalar coupling

A large variety of different heteronuclear shift correlation methods does exist. Not all of them will be treated here, but an attempt will be made to present some experiments in a way that is easy to understand, and gives a useful insight into the fundamentals of correlation methods in general. A summary of experimental aspects of the most convenient shift correlation experiment will be given in section 2.2.4. Because heteronuclear shift correlation has until now been mostly applied to ^{1}H - ^{13}C systems, these nuclei will be used throughout the paragraph although in principle any pair of nuclei with a nuclear spin could be chosen. In the case of ^{1}H - ^{13}C spin system it is most convenient to detect the ^{13}C directly, for reasons explained in section 2.2.5.

2.2.1 The basic experiment

The basic experimental method for the correlation of signals from heteronuclei has first been proposed by Ernst and co-workers, and is described in references 1-3.

In the following discussion of this method an isolated ^{1}H - ^{13}C spin pair with a positive scalar coupling J will be considered. At thermal equilibrium the population of the energy levels can be calculated using the Boltzmann equation. This gives in the high temperature approximation the following relative populations of the energy levels (Fig.2.1):

$$\begin{aligned} P_1 &= 1 - 5p \\ P_2 &= 1 - 3p \\ P_3 &= 1 + 3p \\ P_4 &= 1 + 5p \\ 2p &= \gamma_{^{13}C} H_o \hbar / kT \end{aligned} \qquad [2.1]$$

using $\gamma_{^{13}C} = \gamma_{^{1}H}/4$.

The experiment uses the pulse scheme shown in Fig.2.2. A $\pi/2$ r.f.pulse, applied by the proton transmitter along the x-axis of the proton rotating frame, creates two magnetization components, which will rotate with angular frequencies Ω_{13} and Ω_{24} and amplitudes M_o proportional to $P_3 - P_1$ and $P_4 - P_2$ respectively (Fig.2.3). At a time t_1 after this $\pi/2$ pulse the angles precessed through will be $\Omega_{13}t_1$ and $\Omega_{24}t_1$ (Fig.2.3b), where $\Omega_{13} = \Omega_o + \pi J$ and $\Omega_{24} = \Omega_o - \pi J$, Ω_o denoting the chemical shift frequency of the protons. The second $\pi/2$ pulse to the protons then creates longitudinal magnetization components

$$\begin{aligned} Mz_{13} &= -M_o \cos(\Omega_{13} t_1) \\ Mz_{24} &= -M_o \cos(\Omega_{24} t_1). \end{aligned} \qquad [2.2]$$

As long as no pulse to the ^{13}C spins has been applied the two-spin system can be considered as two one-spin systems, one with the ^{13}C spin in the β-state (levels 1 and 3) and one with the ^{13}C spin in the α-state. The sum of the populations of the energy levels of these two one-spin systems remains unchanged under the pulse sequence applied to the protons:

$$P_1(t_1) + P_3(t_1) = P_1 + P_3 = 2(1 - p) \qquad [2.3a]$$

$$P_2(t_1) + P_4(t_1) = P_2 + P_4 = 2(1 + p) \qquad [2.3b]$$

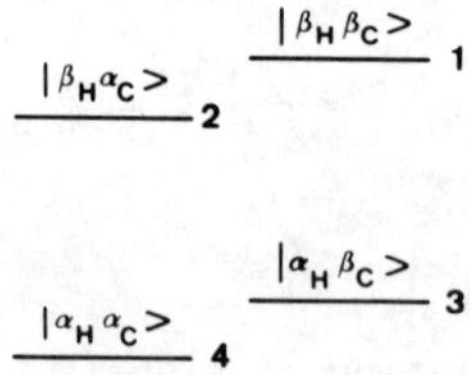

Fig.2.1 Energy level diagram and wave functions of an isolated ^{13}C - ^{1}H spin system.

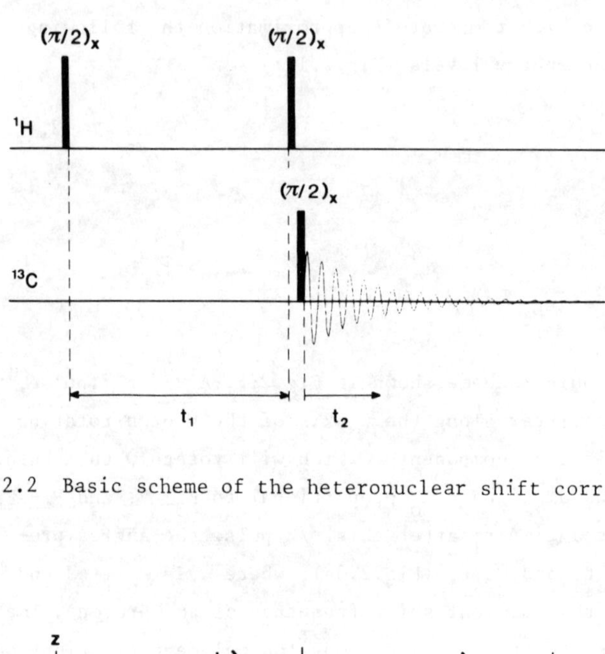

Fig.2.2 Basic scheme of the heteronuclear shift correlation experiment.

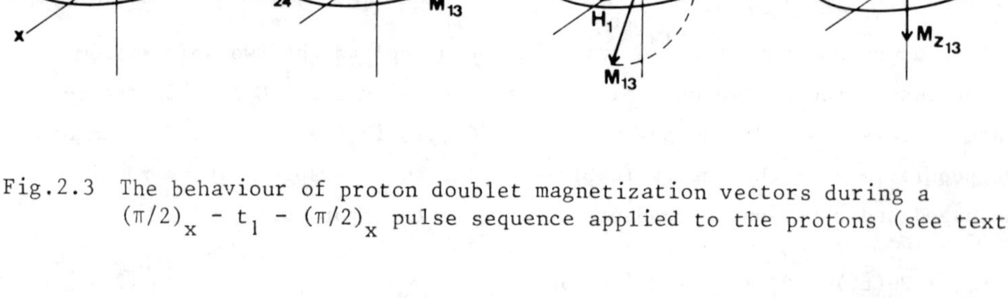

Fig.2.3 The behaviour of proton doublet magnetization vectors during a $(\pi/2)_x - t_1 - (\pi/2)_x$ pulse sequence applied to the protons (see text).

Because the longitudinal magnetization is directly proportional to the difference in occupation of the energy levels, after this $(\pi/2)_x - t_1 - (\pi/2)_x$ sequence applied to the protons the spin state populations will be as follows:

$$
\begin{aligned}
P_1(t_1) &= 1 - p + 4p\cos(\Omega_{13}t_1) \\
P_2(t_1) &= 1 + p + 4p\cos(\Omega_{24}t_1) \\
P_3(t_1) &= 1 - p - 4p\cos(\Omega_{13}t_1) \\
P_4(t_1) &= 1 + p - 4p\cos(\Omega_{24}t_1).
\end{aligned}
\qquad [2.4]
$$

Equation [2.4] shows that the longitudinal magnetization for the two ^{13}C transitions has changed because of the pulse sequence applied to the protons, according to

$$M_{z12} = M_o(^{13}C) + 2M_o(^{13}C)\{\cos(\Omega_{24}t_1) - \cos(\Omega_{13}t_1)\} \qquad [2.5a]$$

$$M_{z34} = M_o(^{13}C) - 2M_o(^{13}C)\{\cos(\Omega_{24}t_1) - \cos(\Omega_{13}t_1)\}. \qquad [2.5b]$$

$M_o(^{13}C)$ is equal to $M_o(^1H)/16$. In the discussions given below, the index ^{13}C will be omitted. As can be seen in Eq.[2.5] the magnitude of the two longitudinal ^{13}C magnetization vectors is modulated (in opposite sense) as a function of t_1 with the proton resonance frequencies. A ^{13}C $\pi/2$ pulse, applied immediately after the second proton pulse (Fig.2.2) will create two transverse ^{13}C magnetization vectors*, which will rotate during time t_2 with angular frequencies Ω_{12} and Ω_{34}. The detected ^{13}C signal is thus described by

$$
\begin{aligned}
s(t_1,t_2) = {} & M_o[1 + 2\{\cos(\Omega_{24}t_1) - \cos(\Omega_{13}t_1)\}]\exp(i\Omega_{12}t_2) + \\
& + M_o[1 - 2\{\cos(\Omega_{24}t_1) - \cos(\Omega_{13}t_1)\}]\exp(i\Omega_{34}t_2).
\end{aligned}
\qquad [2.6]
$$

The real part of the Fourier transform with respect to t_2 of Eq.[2.6] gives two resonance lines in absorption mode at the ^{13}C frequencies Ω_{12} and Ω_{34}, with amplitudes proportional to $1 + 2\{\cos(\Omega_{24}t_1) - \cos(\Omega_{13}t_1)\}$ and $1 - 2\{\cos(\Omega_{24}t_1) - \cos(\Omega_{13}t_1)\}$ respectively. Thus, a second Fourier transform of the data matrix with respect to t_1 will give six resonance lines

---

* Another effect of the ^{13}C $\pi/2$ pulse is that the components of the two proton vectors which are parallel to the x-axis after the second ^1H $(\pi/2)_x$ pulse are converted into zero- and double quantum coherence, as can be shown with a density matrix treatment (section 1.2). If the order of the second proton pulse and the ^{13}C pulse would be reversed, the simple population arguments cannot be used. However, a density matrix calculation shows that the eventual results would remain unchanged.

at

$$(\omega_1,\omega_2) = (0,\Omega_{12}),\ (0,\Omega_{34}),\ (\Omega_{24},\Omega_{12}),\ (\Omega_{24},\Omega_{34}),\ (\Omega_{13},\Omega_{12}),\ (\Omega_{13},\Omega_{34}).$$

Two of these resonances will have negative intensity, corresponding to the minus-signs in Eq.[2.6].

The F_1 and F_2 frequency coordinates in the 2D spectrum are identical to the frequencies as observed in conventional one-dimensional ^1H and ^{13}C spectra. The two axial peaks ($\omega_1 = 0$) originate from the components which are not modulated as a function of t_1. As an example in Fig.2.4 a spectrum of chloroform, obtained with this method, is shown.

2.2.2 Spectral simplification via heteronuclear decoupling

This section describes how spectra can be simplified and sensitivity can be enhanced by decoupling of protons during detection, and apparent decoupling of carbon-13 during the evolution. It will turn out that special precautions have to be taken to avoid mutual cancellation of signals (3,4).

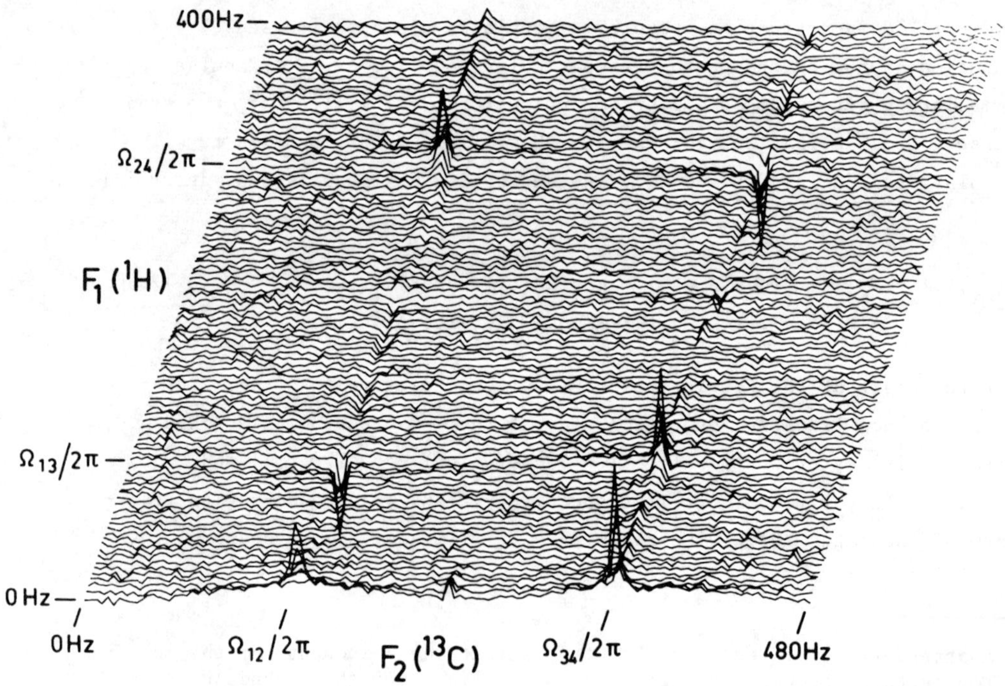

Fig.2.4 Heteronuclear shift correlation spectrum of chloroform ($CHCl_3$) with heteronuclear coupling present in both frequency dimensions. (Recorded at 75 MHz for ^{13}C).

Proton decoupling during detection

As follows from Eq.[2.6] the sum of the two modulated components of the ^{13}C doublet is zero immediately after the $\pi/2$ pulse to the ^{13}C. Thus, if decoupling were started immediately after the $\pi/2$ pulse to the ^{13}C, these modulated signals would cancel. However, if decoupling were started at a delay time $\Delta_2 = (2J)^{-1}$ after the $\pi/2$ detection pulse, the ^{13}C vectors would be parallel again, and a signal at the chemical shift frequency of the carbon-13 would be detected. For arbitrary value of Δ_2 the signal detected is now described by

$$s(t_1,t_2) = 4iM_o\{\cos(\Omega_{24}t_1) - \cos(\Omega_{13}t_1)\}\sin(\pi J\Delta_2)\exp(i\Omega_2\Delta_2)\exp(i\Omega_2 t_2) \quad [2.7]$$
$$+ 2M_o\cos(\pi J\Delta_2)\exp(i\Omega_2\Delta_2)\exp(i\Omega_2 t_2),$$

where Ω_2 denotes the carbon-13 chemical shift frequency. The pulse scheme of this experiment is shown in Fig.2.5. A 2D spectrum of chloroform obtained with this method is shown in Fig.2.6. The measuring time and digital filtering applied were identical to those for the spectrum of Fig.2.4. As expected, the spectrum is simplified and the sensitivity is improved compared with the spectrum of Fig.2.4. Immediately after the $\pi/2$ ^{13}C pulse the unmodulated doublet components are parallel (Eq.[2.6]) so a time $\Delta_2 = (2J)^{-1}$ later they will be antiparallel. Starting the decoupling at this moment will cause cancellation of these signals so that no axial peaks ($\omega_1 = 0$) will be obtained in this case (Eq.[2.7]). As can be seen in Fig.2.6 the delay time Δ_2 did not exactly satisfy this condition and a small axial peak does show up.

Carbon-13 decoupling during evolution

A further sensitivity enhancement and simplification of the spectrum can be obtained if the carbon-13 is decoupled during the evolution period, so that the proton doublet coalesces at the chemical shift frequency. If the ^{13}C decoupling were applied during the whole interval between the two $\pi/2$ proton pulses, no magnetization would be transferred, since the two proton vectors remain parallel during the entire evolution period, and the term in Eq.[2.7] which causes the modulation $\{\cos(\Omega_{24}t_1) - \cos(\Omega_{13}t_1)\}$ would be zero. However, as will be shown below, if a delay $\Delta_1 = (2J)^{-1}$ is inserted during which no ^{13}C decoupling is applied (Fig.2.7), the magnetization vectors of the proton doublet will be opposite just before the second $\pi/2$ pulse, and transfer of magnetization will be optimized.

At the end of the period during which the ^{13}C decoupling is applied (time t_1 in Fig.2.7), both proton doublet components have precessed through

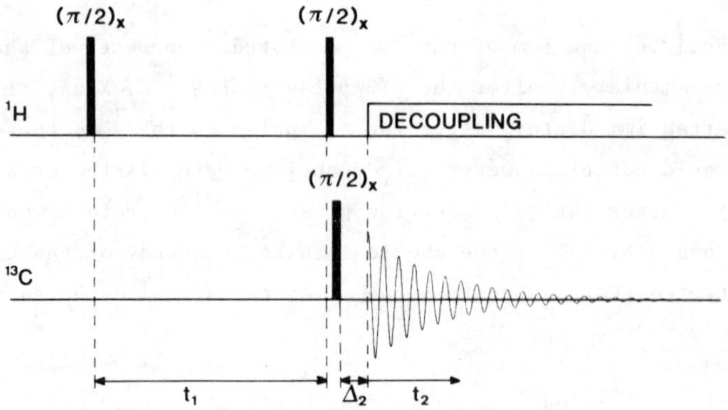

Fig.2.5 Experimental scheme for heteronuclear shift correlation with proton-decoupled acquisition.

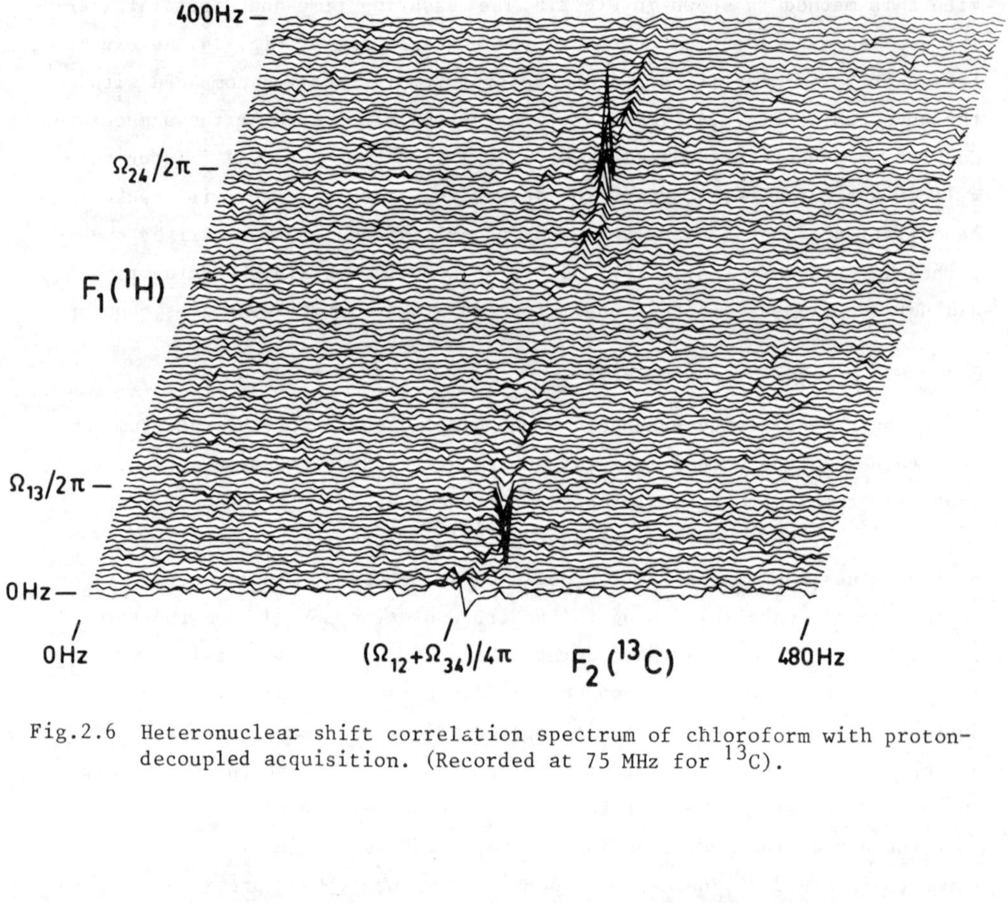

Fig.2.6 Heteronuclear shift correlation spectrum of chloroform with proton-decoupled acquisition. (Recorded at 75 MHz for ^{13}C).

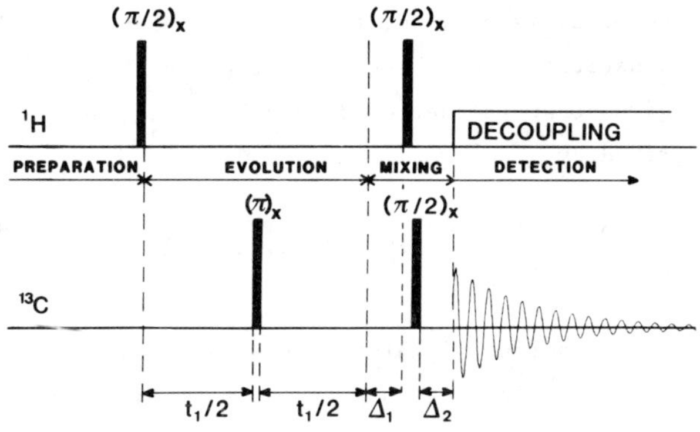

Fig.2.7 Scheme of the heteronuclear shift correlation experiment with ^{13}C decoupling during evolution and ^{1}H decoupling during detection.

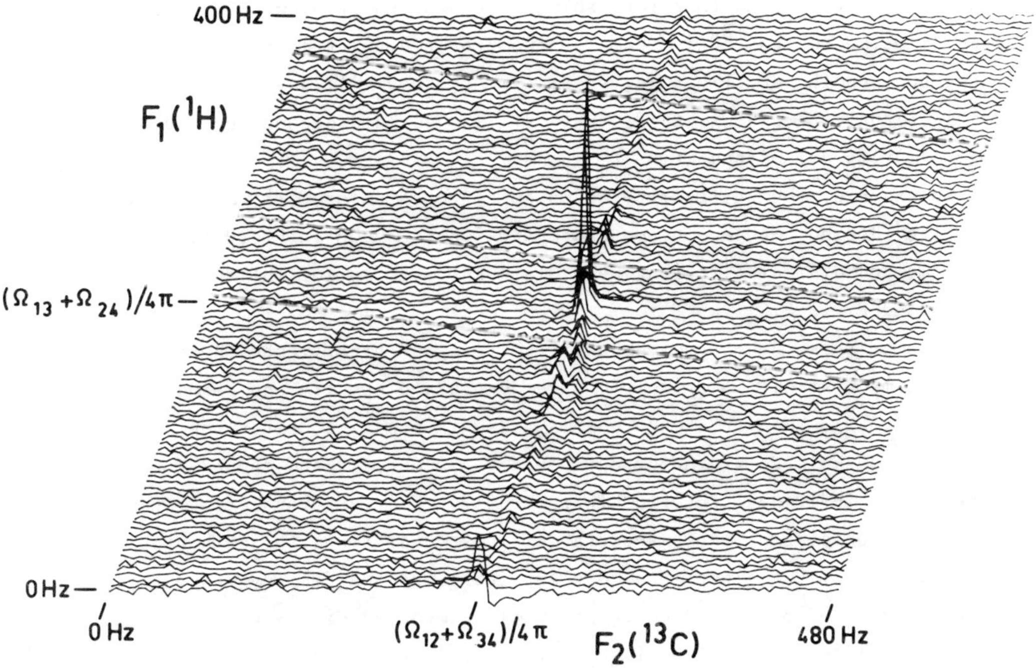

Fig.2.8 Phase-sensitive display of the chemical shift correlation spectrum of chloroform, decoupled along both axes. (Recorded at 75 MHz for ^{13}C).

angles $\Omega_1 t_1$, where Ω_1 is the proton chemical shift frequency. A time Δ_1 later, during which no ^{13}C decoupling is applied, the angles precessed through will be $\Omega_1 t_1 + (\Omega_1 \pm J/2)\Delta_1$. Substitution of these two values in Eq.[2.7] (instead of $\Omega_{24} t_1$ and $\Omega_{13} t_1$) gives that the detected signal in the case of an arbitrary value Δ_1 is described by

$$s(t_1,t_2) = 4iM_o \sin(\Omega_1 t_1 + \Omega_1 \Delta_1)\sin(\pi J \Delta_1)\sin(\pi J \Delta_2)\exp\{i\Omega_2(t_2 + \Delta_2)\} + \\ + 2M_o \cos(\pi J \Delta_2)\exp\{i\Omega_2(t_2 + \Delta_2)\}. \qquad [2.8]$$

As can be seen from this equation, maximum signal is obtained for a length of Δ_1 equal to $(2J)^{-1}$. This means that in this case transfer of magnetization from ^1H to ^{13}C is optimized.

A simple way of apparent decoupling the ^{13}C during the time t_1 is to apply one single ^{13}C π pulse just in the middle of the evolution period (3,4). This requires much less power than broad-band noise decoupling of the wide carbon-13 spectrum, and therefore solves the heating problems too. The effect of the π pulse to the ^{13}C is that the two proton doublet components interchange frequencies at time $t_1/2$, so at time t_1 they will have rotated an identical angle $(\Omega_{13} + \Omega_{24})t_1/2$, just as if broad-band decoupling had been applied during all of that time t_1.

As an example, the shift correlation spectrum of chloroform obtained with decoupling during both evolution and detection is shown in Fig.2.8. Measuring time and digital filtering were again the same as for the spectra of Fig.2.4 and Fig.2.6.

Because in these heteronuclear shift correlation experiments there is amplitude modulation, in principle pure absorption spectra can be obtained (section 1.5.2). However, two problems arise in practice: first the delays Δ_1 and Δ_2 cause large frequency dependent phase errors, and second the noise in the individual spectra after the first Fourier transformation makes interactive phasing of these spectra difficult. Solutions for both problems do exist (4,5,6) but since in a more sophisticated and in many respects better version of the shift correlation experiment as discussed in the next section, amplitude modulation is converted into phase modulation, the problem of phasing becomes irrelevant.

CH_2- and CH_3-groups

Until now, attention has only been paid to isolated ^{13}C - ^1H spin systems. As derived in appendix II similar results can be obtained in the case of $^{13}CH_2$- and $^{13}CH_3$ groups. In the case of decoupling during both evolution and detection periods, the same procedure as described in this section can be used. The only difference turns out to be the optimum choice for the length of Δ_2, while the optimum choice for Δ_1 remains $(2J)^{-1}$. In Fig.2.9 the detected signal amplitude as a function of the length of Δ_2 is shown for ^{13}CH-, $^{13}CH_2$- and $^{13}CH_3$ groups. A choice $\Delta_2 = 0.3\ J^{-1}$ gives good signal amplitude in all three cases.

2.2.3 Separation of positive and negative proton frequencies

Because of the amplitude modulation present in the experiments described previously, it is impossible to distinguish between positive and negative modulation frequencies in a simple way (section 1.5.2). Hence in these experiments care has to be taken that the proton transmitter frequency is either at lower or higher frequency than any of the proton resonances of interest. This implies that the sampling frequency along the t_1-axis has to be at least twice the spectral width of the ^1H region of interest to avoid folding. To decouple the ^1H region with noise irradiation either the proton synthesizer has to be switched to the frequency of the centre of the proton region, or a larger decoupling power must be used. A modified sequence converts amplitude into phase modulation, making it possible to put the proton transmitter permanently in the centre of the ^1H spectrum, decreasing the decoupling power required and halving both the sampling frequency along the t_1-axis and the data storage space required (7). The actual pulse scheme is the same as shown in Fig.2.7, but now the experiment is performed in two stages (Fig.2.10).

The first stage is an experiment with the phase of the second $\pi/2$ proton pulse along the x-axis, and the second with the pulse applied along the

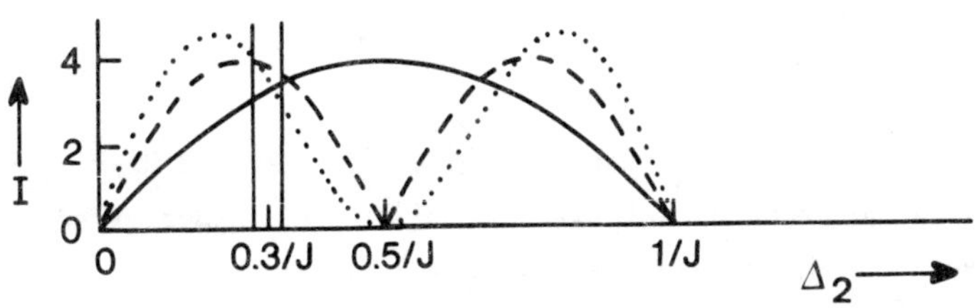

Fig.2.9 The Δ_2 dependence of the magnitude I of that part of the magnetization in a decoupled shift correlation experiment which is modulated as a function of t_1 for a ^{13}CH group (drawn line), a $^{13}CH_2$ group (broken line) and a $^{13}CH_3$ group (dotted line). Magnitude 1 corresponds to the total ^{13}C magnetization in thermal equilibrium (without Overhauser enhancement).

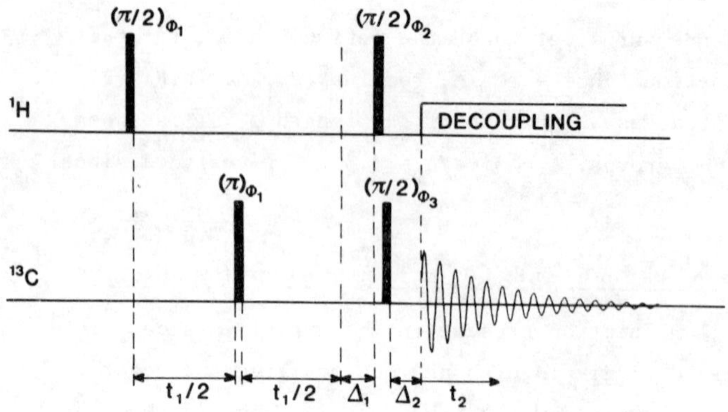

Fig.2.10 Scheme of the heteronuclear shift correlation experiment with decoupling during evolution and detection. In order to distinguish between positive and negative proton modulation frequencies and to suppress axial peaks the phases of the r.f.pulses and the phase Ψ of the receiver are cycled as indicated in Table 2.1.

y-axis of the proton rotating frame and the receiver reference phase incremented by 90°. All other phases are identical in the two stages. For an isolated ^{13}C - ^{1}H spin pair the signals detected in the two stages are described by

$$s(t_1,t_2) = C_1\sin(\Omega_1 t_1 + \Omega_1\Delta_1)\exp(i\Omega_2 t_2) + C_2\exp(i\Omega_2 t_2) \qquad [2.9a]$$

$$s(t_1,t_2) = -iC_1\cos(\Omega_1 t_1 + \Omega_1\Delta_1)\exp(i\Omega_2 t_2) - iC_2\exp(i\Omega_2 t_2), \qquad [2.9b]$$

where Ω_1 and Ω_2 are again the ^{1}H and ^{13}C angular chemical shift frequencies. The constant C_1 is equal to $4iM_o\sin(\pi J\Delta_1)\sin(\pi J\Delta_2)\exp(i\Omega_2\Delta_2)$ and C_2 is equal to $2M_o\cos(\pi J\Delta_2)\exp(i\Omega_2\Delta_2)$ (Compare Eq.[2.8]).

Direct addition of the two signals of Eq.[2.9] gives

$$s(t_1,t_2) = C_1'\exp(i\Omega_1 t_1)\exp(i\Omega_2 t_2) + C_2'\exp(i\Omega_2 t_2) \qquad [2.10]$$

with $C_1' = -iC_1\exp(i\Omega_1\Delta_1)$ and $C_2' = \sqrt{2}\, C_2\exp(-i\pi/4)$.

The first term at the right hand side of Eq.[2.10] is a pure phase-modulated signal on which a complex Fourier transformation can be performed along the t_1-axis, distinguishing between positive and negative values for Ω_1. Because of the phase modulation a phase-twisted line shape will be obtained (section 1.5.1), and an absolute value calculation is usually made before display (section 1.5.6).

Since in the first term at the right hand side of Eq.[2.10] both frequencies appear with the same sign in front, this signal represents the antiecho component (section 1.3)*, subtraction of the results of the two stages would have given opposite signs for Ω_1 and Ω_2, representing the coherence transfer echo component. (Note that subtraction of the signal of Eq. [2.9b] is identical to an extra increment of the receiver reference phase by 180°). This latter component is usually preferred because of the higher sensitivity in the case of an inhomogeneous magnetic field.

Suppression of axial peaks

The component in Eq. [2.10] with the constant C_2' is not modulated as a function of t_1 and will thus cause axial peaks at $(\omega_1,\omega_2) = (0,\Omega_2)$. These signals are generally unwanted and can easily be eliminated, as shown below.

If in the same sequence which gave the signal of Eq. [2.9] the second proton pulse is applied along the -x-axis instead of the +x-axis, this has the same effect as if both proton magnetization vectors had rotated an extra angle π about the z-axis in the original experiment. This then gives for the signal

$$s(t_1,t_2) = -C_1\sin(\Omega_1 t_1 + \Omega_1\Delta_1)\exp(i\Omega_2 t_2) + C_2\exp(i\Omega_2 t_2). \qquad [2.11]$$

Subtracting this signal from the signal of Eq. [2.9a] causes cancellation of the unmodulated component. In a similar way the unmodulated component in Eq. [2.9b] can be cancelled by subtracting a signal from an experiment with the second proton pulse along the -y-axis.

A complete scheme for the phases of pulses and receiver in the four different stages of the experiment, necessary to separate positive and negative modulation frequencies and to suppress axial peaks, is given in Table 2.1.

Table 2.1 Phases ϕ_1, ϕ_2 and ϕ_3 of the radiofrequency pulses in Fig.2.10, and phase Ψ of the receiver through which the experiment has to be cycled for each value of t_1 for detecting either the antiecho (Ψ_a) or the coherence transfer echo (Ψ_e) selectively and for suppression of axial peaks.

Experiment number	ϕ_1	ϕ_2	ϕ_3	Ψ_a	Ψ_e
1	x	x	x	x	x
2	x	y	x	y	-y
3	x	-x	x	-x	-x
4	x	-y	x	-y	y

* Since all equations derived until now are also valid in the laboratory frame, this is consistent with the definition given in section 1.3.

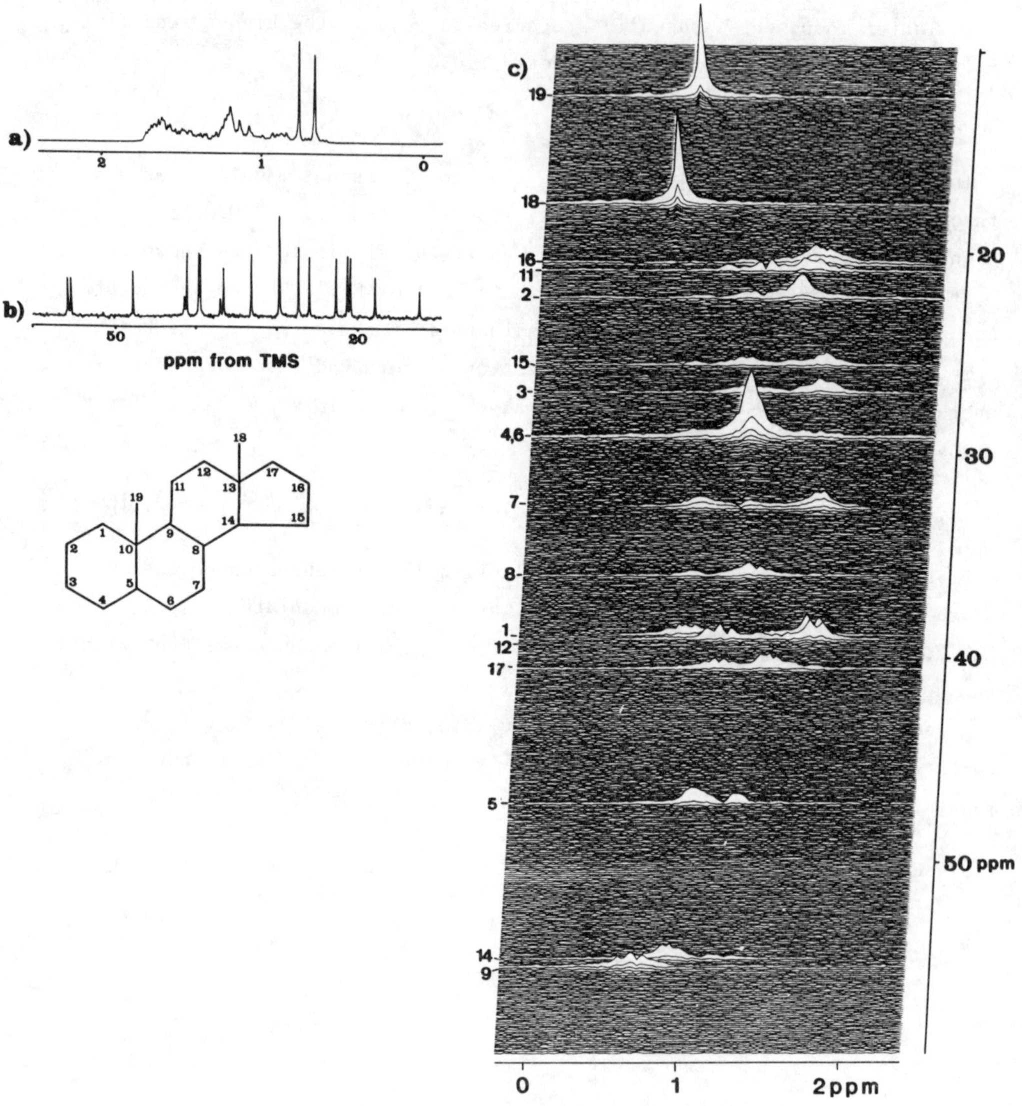

Fig.2.11 (a) The conventional proton spectrum of 5 α-androstane recorded at 200 MHz.
(b) The conventional broadband proton-decoupled ^{13}C spectrum recorded at 50 MHz.
(c) The heteronuclear shift correlation spectrum.
The resonances are broadened along the F_1-axis by homonuclear proton coupling. From ref.(7).

A typical example of a spectrum obtained with this method is shown in Fig. 2.11.

2.2.4 The experimental procedure

The heteronuclear shift correlation experiment appears to be very useful in solving practical problems. Because the variety in experimental methods and the number of settings to be chosen is large, in this section a strategy will be given for obtaining a near optimum spectrum.

The pulse sequence of Fig.2.10 should be used with the phase cycling of pulses and receiver as indicated in Table 2.1. This sequence gives the simplest eventual spectrum, heteronuclear decoupled along both frequency axes, and requires less data storage space then the other sequences. Positive and negative proton frequencies are distinguished and axial peaks are effectively suppressed (7).

In order to cycle out errors in the quadrature detection system, the whole four-step experiment can be repeated four times, incrementing all phases of pulses and receiver each time by 90° (7), analogous to the cyclops experiment proposed by Hoult (8).

The following points demand attention:

Delay Δ_1

Because most direct ^{13}C - ^{1}H coupling constants are in the range 125-170 Hz, a delay Δ_1 equal to 3.3 millisec. will be close to optimum in almost all cases.

Delay Δ_2

As discussed in section 2.2.2 the optimal length of Δ_2 is different for CH, CH_2 and CH_3 groups. A time of 2 millisec. will usually show all correlations with sufficient intensity.

Experiment delay time T

Because the magnetization is transferred from protons to carbon-13, the optimum length of the delay time T between the end of acquisition and the first proton pulse of the next sequence depends on the longitudinal relaxation time T_1 of the protons, rather than on the usually longer T_1's of the carbon-13. A delay time $T = 1.3T_1$, where T_1 is the longitudinal relaxation time of the protons gives optimal sensitivity (9). Care has to be taken that during this delay time T no r.f.power from the ^{1}H transmitter in the frequency region of the proton spectrum leaks through to the transmitter coil, because this will cause (partial) saturation of the proton transitions, degrading sensitivity.

Acquisition times

In cases of poor sensitivity, the sampling time along the t_2-axis should at least be equal to the decay constant T_2 of the ^{13}C signal to avoid sensitivity loss. The length of the acquisition time along the t_1-axis depends on the ^1H resolution required, the amount of data storage space available, and the sensitivity. An acquisition time along the t_1-axis of 100 millisec. is usually sufficient.

Acquisition frequency along the t_1-axis

The minimum acquisition frequency along the t_1-axis is twice the largest offset frequency from the proton transmitter to a ^1H resonance of interest, to avoid folding. The ^1H transmitter frequency has to be set at the centre of the ^1H region of interest, to minimize the acquisition frequency required.

Calibration of the proton pulses

The proton pulses are usually applied using the proton decoupler in the coherent mode for a short period of time. The calibration of the width of the proton pulses is not very critical, since, as can be derived, the amplitude of the modulation is approximately proportional to $\sin^2(\alpha)$ if α proton pulses are applied. As the strength of the decoupler field cannot be determined directly via a 180° pulse calibration as used for calibrating the observe pulses, an indirect approach has to be chosen. A convenient method for calibrating flip angle α to $\pi/2$ is to apply a $\pi/2(^{13}C) - \tau - \alpha(^1H)$ pulse sequence to a simple compound like e.g. chloroform, and setting the delay τ carefully to $1/2J$. If $\alpha = \pi/2$ the detected carbon-13 signal will be zero.

Shimming the magnet

Because a large sample diameter is usually used for the detection of ^{13}C, proton spinning side bands will easily occur, since the effect of field inhomogeneity is four times worse for protons than for carbon-13. Careful shimming of the sample in the non-spinning mode is therefore important to limit the size of these peaks.

2.2.5 Direct detection of the protons

As mentioned in the introduction of paragraph 2.2, the ^{13}C signal is usually detected during acquisition. In principle however, the pulse sequences applied to protons and carbon-13 could be interchanged. Since signal amplitude is proportional to the population difference, to the magnetogyric ratio, and to the frequency of the transitions one might expect a gain in sensitivity equal to $(\gamma_{^1H}/\gamma_{^{13}C})^3 = 64$ in this case. In many cases, however, this factor will be much smaller, and the experimental problems will be more difficult if the protons are detected directly. This is due to the following reasons:

(a) The much stronger signal from protons not coupled to ^{13}C has to be suppressed. The complete suppression of these 200 times stronger signals is a difficult problem.

(b) Broad-band decoupling of ^{13}C during acquisition is in practice impossible because of the enormous decoupling power required.

(c) In the experiment with ^{13}C detection during t_2, proton population differences are transferred to ^{13}C, increasing signal amplitude by a factor of four. In the experiment with ^{1}H detection during t_2, ^{13}C population differences are transferred to ^{1}H, decreasing signal amplitude by a factor of four. Bodenhausen and Ruben (10) proposed to transfer, in the case of proton detection during t_2, proton populations to the rare nucleus (^{13}C or ^{15}N), prior to the start of the experiment. In this latter case the decrease in signal amplitude by a factor of four does not occur.

(d) Because of the much larger spectral width of ^{13}C, a much higher sampling frequency along the t_1-axis and hence a much larger number of t_1 increments is required, lengthening the minimum measuring time.

(e) Since the decay rate of the proton resonances, broadened by unresolved proton-proton coupling is usually much larger than those of proton decoupled ^{13}C resonances, a further loss in sensitivity occurs.

(f) The experimental delay time (section 2.2.4) will be determined by the often much longer longitudinal relaxation time of the rare nucleus, decreasing the repetition rate of the sequence.

These considerations make it unattractive to detect the protons directly in the heteronuclear shift correlation experiment in the case of $^{13}C - {}^{1}H$ spin systems. However, the possible gain in sensitivity can in the case of $^{15}N - {}^{1}H$ spin systems ($\gamma_{^1H}/\gamma_{^{15}N} = 10$) be a reason to detect the protons directly (10).

However, since the abundance of ^{15}N is only 0.37% and many interesting nitrogen containing compounds as peptides are only soluble in water, the dynamic range (11) and the suppression mentioned under point (a) are even harder problems. Until now these experimental problems have not been solved completely.

2.2.6 Indirect observation of homonuclear couplings

In this section a two-dimensional experiment will be described which gives on cross-sections taken parallel to the F_1-axis and cutting the F_2-axis at a chemical shift frequency of a ^{13}C, the proton multiplet structure of the proton directly coupled to the ^{13}C.

These results are analogous to those obtained with J-spectroscopy (chapter 3), but since the experimental method is quite similar to the heteronuclear shift correlation scheme, the experiment will be treated in this chapter.

As can be seen in Fig.2.11 in the heteronuclear correlation spectra the proton-proton coupling structure remains present along the F_1-axis of the 2D spectrum. Usually this structure is not resolved because of poor digitization. Since these proton resonances are essentially the same as the carbon-13 satellites in conventional proton spectra, they will have the same multiplet structure provided that these satellites are first order, in principle allowing a determination of the proton-proton coupling constants. Indirect determination of proton-proton coupling constants can be important where this information cannot be extracted from the conventional ^1H spectrum or from the two-dimensional J-spectrum (section 3.3), because of too severe overlap.

There are three main problems with the measurement of proton-proton couplings from a normal heteronuclear shift correlation experiment (section 2.2.4):

(a) Because the proton multiplet has to be resolved, signal energy is spread over several resonances along the F_1-axis and sensitivity will be poor. Since sensitivity enhancement filters cause line-broadening and therefore cannot be used, this problem becomes even worse.
(b) Good discretization, necessary to resolve the multiplet structure, generally requires a large number of t_1 increments and thus a large data matrix.
(c) The line width along the F_1-axis will be broadened by extra inhomogeneity effects due to the use of a large sample diameter, required for sufficient sensitivity.

The last two problems can be overcome by the use of a slightly modified sequence, shown in Fig.2.12. The idea is to apply a π pulse to the protons in the centre of the evolution period (12), removing the effects of proton chemical shift and heteronuclear coupling during the time t_1. Thus, the transferred signal is only modulated by the homonuclear proton-proton coupling, which is not affected by the π pulse (section 3.3). Since in the case of weak coupling the proton multiplet is symmetric, it is not necessary to distinguish between positive and negative modulation frequencies. Axial peaks can be suppressed by performing the experiment in two stages and co-adding the results: The first stage with the second ^{1}H $\pi/2$ pulse applied along the y-axis and the ^{13}C $\pi/2$ pulse applied along the x-axis, and the second stage with the pulses applied along the -y- and -x-axis respectively (Fig.2.12). The effect of the two π pulses in the centre of the delay period Δ_1 is to remove the effect of large phase errors along the F_1-axis, by eliminating the effects of chemical shifts and retaining the (constant) effect of heteronuclear coupling (section 3.1) necessary for the magnetization transfer from protons to carbons (5). This makes it possible to obtain a two-dimensional absorption spectrum (13).

The spectral width along the F_1-axis in this modified experiment is now equal to the proton multiplet width, and hence a low sampling frequency of about 30 Hz is usually sufficient. Therefore, only a relatively small data matrix is needed. Since the inhomogeneity broadening is removed by the π pulse (section 3.1), in principle natural line widths for the proton resonances are obtained.

This experiment is closely related to homonuclear J-spectroscopy as described in section 3.3, and is generally referred to as indirect J-spectroscopy. As is clear, proton multiplets with exactly the same chemical shift can still be separated with this method if the directly coupled carbons have different shifts.

As an illustration of the technique, in Fig.2.13 the indirect J-spectrum of sucrose is shown.

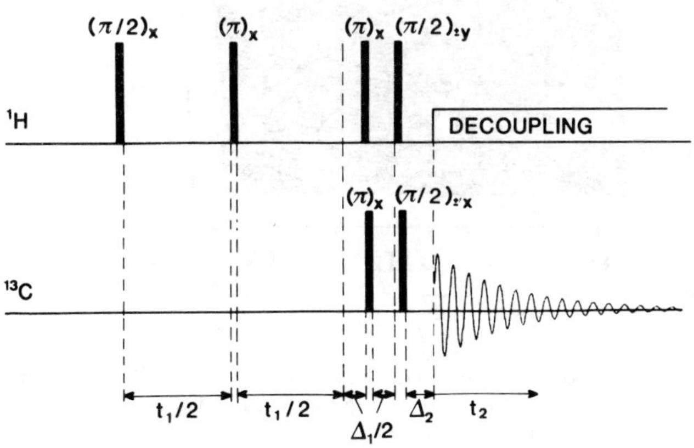

Fig.2.12 Pulse scheme for the indirect detection of homonuclear proton coupling.

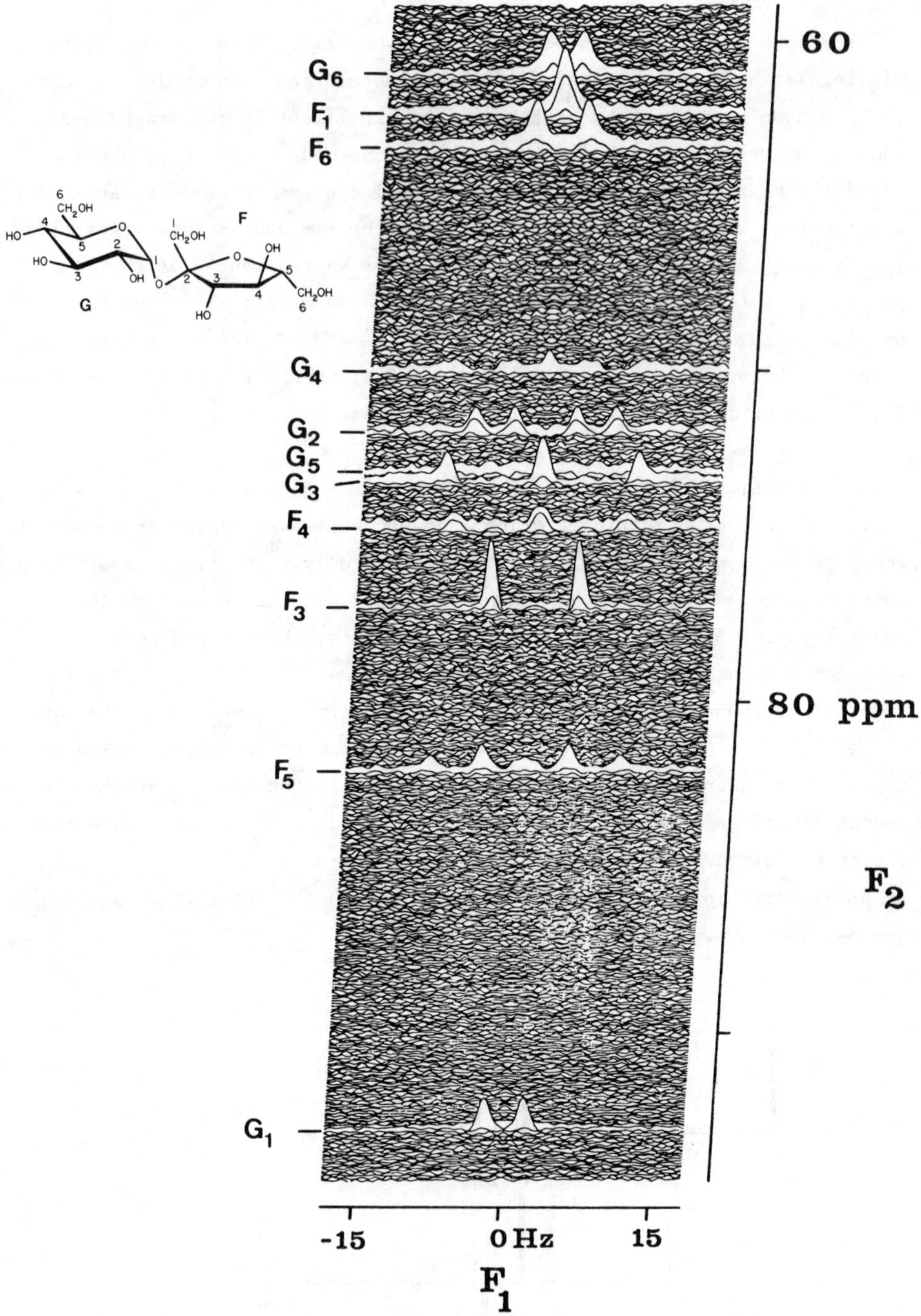

Fig.2.13 Indirect J-spectrum of sucrose recorded at 50 MHz ^{13}C frequency. In the F_1 dimension the homonuclear proton multiplet structure is present. In the F_2 dimension the frequency of a resonance line is equal to the chemical shift frequency of the directly coupled carbon-13. The traces showing the multiplet structure of the protons coupled to F_4, F_5, G_4 and G_5 are distorted due to strong coupling in the proton spectrum. From ref.(13).

2.3 Correlation of chemical shifts through homonuclear scalar coupling

The very first two-dimensional experiment, proposed by Jeener in 1971 (14), was a homonuclear shift correlation experiment. The experiment was later analyzed in detail by Aue et al. (15). Initially the experiment was proposed as an alternative to homonuclear double resonance (16), i.e. for determining the connectivity★ patterns in homonuclear spin systems. As has subsequently been shown (17) many analogies between these two types of experiments exist. Quite a few modifications on Jeener's basic pulse scheme have been proposed recently, enlarging the applicability of the experiment. Modifications were proposed for the determination of relative signs of coupling constants, detection of very weak couplings, and for removing the homonuclear splittings along the F_1-axis of the two-dimensional spectrum (17). Other modifications were proposed with the purpose of limiting the size of the data matrix required (18,19,20).

In this paragraph the basic pulse scheme and the most important variations mentioned above, will be discussed. For a detailed description of the theory of the experiment the reader is referred to ref.(15). In section 2.2.3 an attempt will be made to explain the basic mechanism of the experiment via a simple physical picture.

2.3.1 The basic experiment

In its simple form Jeener's experiment employs the radiofrequency pulse sequence

$$T - (\pi/2)_x - t_1 - (\pi/2)_x - t_2,$$

where T is the experiment delay time between the end of acquisition and the start of the next sequence, applied to a system of homonuclear coupled spins, usually protons (Fig.2.14). The first $\pi/2$ pulse along the x-axis, applied to a spin system in thermal equilibrium, creates transverse magnetization for all allowed transitions mn, where $|M_m - M_n| = 1$, and in which only one spin changes its polarization. M_m denotes the total magnetic quantum number of state m. All transverse magnetization components precess during the evolution period with their characteristic angular frequency Ω_{mn} around the z-axis.

★ According to the old definition of Anderson and Freeman (16), two transitions are connected if they share a common energy level. In this section the word transition rather than coherence will be used since it agrees better with literature.

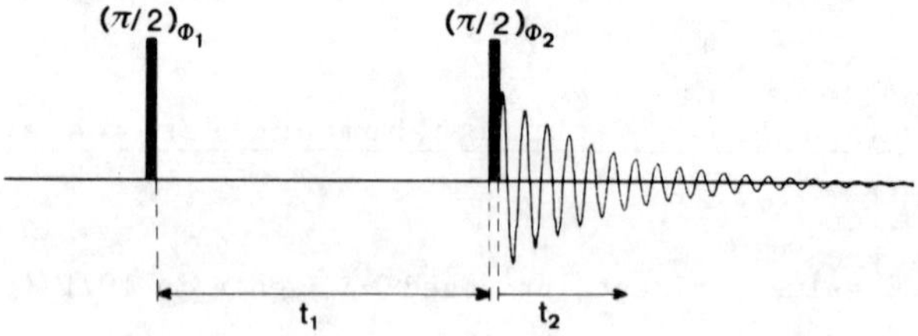

Fig.2.14 Pulse scheme for the homonuclear shift correlation scheme. In the original scheme of Jeener, the pulses are applied along the x-axis (ϕ_1,ϕ_2= x) and the receiver phase (Ψ) is constant. For separating the positive and negative modulation frequencies (quadrature Jeener) the phases ϕ_1,ϕ_2 and Ψ are cycled according to Table 2.2.

At time t_1 the second $(\pi/2)_x$ pulse causes a redistribution of the coherence corresponding to magnetization component mn among all coherences possible in the spin system considered (Eq. [1.27]). As derived by Aue et al. (15), the amount of coherence that is transferred by this mixing pulse from coherence mn to single quantum coherence* kl is proportional to $\sin(\Omega_{mn}t_1 + \phi_{kl,mn})$ (see also section 1.2.3, Eq. [1.30]). The phase constant ϕ is a multiple of 90° and depends on the connectivity of transitions mn and kl (21,15). In the new, extended definition by Aue et al. (15) any pair of transitions in a spin system of N non-equivalent coupled spin-½ nuclei is either parallel, or progressively or regressively connected. If mn and kl are transitions in the same multiplet, the transitions are said to be parallel and the phase constant $\phi_{kl,mn}$ equals 0°. If a spin A flips in transition mn and spin B flips in transition kl, transitions mn and kl are regressively connected if spin B in the states m and n has the same polarity as spin A in states k and l; in the case of opposite polarity the transitions are progressively connected. Hence, if the two transitions belong to different multiplets, they are either progressively or regressively connected. The phase constant $\phi_{kl,mn}$ will then be -90° or $+90^o$ (15,22).

A magnetization component detected during t_2, rotating with angular frequency Ω_{kl}, then originates from all transverse magnetization components present during t_1 and belonging to the same spin system:

$$s_{kl}(t_1,t_2) = \sum_{mn} C_{kl,mn}\sin(\Omega_{mn}t_1 + \phi_{kl,mn})\exp(i\Omega_{kl}t_2 + i\Psi_{kl,mn}), \qquad [2.12]$$

where the summation extends over all coherences present after the first $\pi/2$ pulse. $\Psi_{kl,mn}$ is in analogy with $\phi_{kl,mn}$ a phase constant which is equal to 0°,

---

* As will be discussed in chapter 4 the second $\pi/2$ pulse also creates multiple quantum coherence.

-90° or 90°, depending on whether transition kl and mn are parallel, progressively or regressivly connected. $C_{kl,mn}$ is a constant which equals zero if kl and mn belong to a different symmetry group (23) in the case of equivalent nuclei in the spin system. In the case of an N spin-$\frac{1}{2}$ system with no overlapping resonances and no equivalent nuclei all constants C will have the same value.

In general all detected resonances can thus be modulated as a function of t_1 by all frequencies generated after the first pulse. A two-dimensional Fourier transformation will then produce a spectrum with resonance lines at $(\Omega_{mn},\Omega_{kl})$ if kl and mn are transitions in the spin system of a single molecule. Hence, if two nuclei A and X are coupled, two sets of cross peaks will occur in the two-dimensional spectrum centered at (δ_A,δ_X) and (δ_X,δ_A) (Fig. 2.15). Because of the phase constants $\phi_{kl,mn}$ and $\Psi_{kl,mn}$ in Eq. [2.12] the phases of the different resonances in the 2D spectrum are different; if e.g. the diagonal peaks are in the two-dimensional dispersion mode, the cross peaks will be in positive or negative absorption mode. Longitudinal magnetization present at the end of the evolution period gives rise to signals during t_2 which are not modulated as a function of t_1 and therefore cause axial peaks (section 1.2.3) at $(0,\delta_A \pm J/2)$ and $(0,\delta_X \pm J/2)$, not shown in Fig.2.15.

2.3.2 Improvements of the basic experiment

Since in the basic experiment as discussed in the previous section, the detected signals are modulated in amplitude rather than in phase, it cannot be determined whether the modulation frequency is positive or negative (sec-

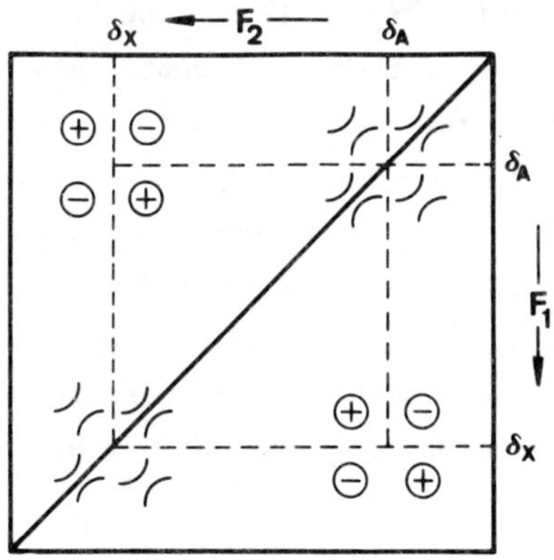

Fig.2.15 Schematic diagram of the Jeener spectrum of an AX spin system, showing the diagonal peaks which have a dispersion shape in both dimensions, and the cross peaks (circles) which have an absorption shape in both dimensions but which alternate in sense (+,-). Diagonal peaks are centred at the coordinates (δ_A,δ_A) and (δ_X,δ_X) while cross peaks are centred at (δ_A,δ_X) and (δ_X,δ_A). Cross peaks indicate spin coupling between A and X. From ref.(17).

tion 1.5.2). Therefore, to be sure, the spectrum has to be at one side of the transmitter frequency, inferring single detection during t_2 and a twice as large offset for the resonance line at the far end of the spectrum, compared with a quadrature detection experiment. If the transmitter frequency is set to the centre of spectrum, and quadrature detection is used during t_2, the highest modulation frequency would be twice as low, requiring a sampling frequency along the t_1-axis twice as low, and therefore a data matrix twice as small★. The problem remaining now is to distinguish between positive and negative modulation frequencies. In this section a modification will be discussed which converts the amplitude modulation into phase modulation and removes the axial peaks. Furthermore, attention will be paid to improving the appearance of the spectrum by the use of digital filtering.

In a similar way to that described in section 2.2.3 for the heteronuclear case, the sign of the modulation frequency can be determined by performing additional experiments $(T - (\pi/2)_x - t_1 - (\pi/2)_y - t_2)$ (19) for which the signal is described by

$$s_{kl}(t_1,t_2) = \sum_{mn} C_{kl,mn}\cos(\Omega_{mn}t_1 + \phi_{kl,mn})\exp(i\Omega_{kl}t_2 + i\Psi_{kl,mn} - i\pi/2), \quad [2.13]$$

where $C_{kl,mn}$, $\phi_{kl,mn}$ and $\Psi_{kl,mn}$ have the same values as in Eq. [2.12].
The difference of the two signals of Eq. [2.12] and [2.13] gives

$$s_{kl}(t_1,t_2) = \sum_{mn} iC_{kl,mn}\exp(-i\Omega_{mn}t_1 - i\phi_{1k,mn})\exp(i\Omega_{kl}t_2 + i\Psi_{kl,mn}). \quad [2.14]$$

Eq. [2.14] represents a signal, phase-modulated as a function of t_1, from which the sign of the modulation frequency is automatically determined in a two-dimensional complex Fourier transformation. Since in Eq. [2.14] the frequency terms Ω_{kl} and Ω_{mn} have opposite signs in front of them, this represents the coherence transfer echo part of the signal (section 1.3 and footnote section 2.2.3). Resonances which do not change their frequency during the $\pi/2$ mixing pulse ($\omega_{kl} = \omega_{mn}$) will give rise to diagonal peaks in the two-dimensional spectrum. The sum of the two signals of Eq. [2.12] and [2.13] would have given identical signs for the two frequencies Ω_{kl} and Ω_{mn} in Eq. [2.14], representing the antiecho component (section 1.3). Usually the coherence transfer echo component is preferred because of the (slightly) higher sensitivity and

★ Note that the signal after the first Fourier transfomation $S(t_1,\omega_2)$ consists of complex data points, both in the case of single and quadrature detection during t_2. Thus for the input of the second Fourier transformation a complex signal is always available.

better resolution.

In order to eliminate the axial peaks at $\omega_1 = 0$, due to longitudinal relaxation during t_1, (section 1.2.3) a further phase cycling is required (19,20) as indicated in Table 2.2. Thus at least four experiments are required for each value of t_1 in order to distinguish positive and negative modulation frequencies and to eliminate the axial peaks.

Another solution, achieving the same goals but requiring only one experiment for each value of t_1, is the use of pulsed field gradients. The experimental scheme is shown in Fig.2.16. The two identical pulsed field gradients, just before and after the mixing pulse, ensure that during the detection period only the coherence transfer echo component can be observed (24). The second field gradient pulse defocusses all transverse magnetization which originates from magnetization that was longitudinal just before the mixing pulse, and which would have caused the axial signals. The sensitivity of the experiment is not improved compared with the phase cycle version of Table 2.2, but the minimum measuring time is shortend by a factor of four. In practice good separation of positive and negative modulation frequencies and good suppression of axial peaks will only be obtained if static magnetic field inhomogeneity is very small compared with the gradients applied during the field gradient pulse.

In order to eliminate artefacts due to errors in the quadrature detection system, both the four-step experiment and the pulsed field gradient experiment can be repeated four times, incrementing all phases of pulses and receiver by 90^0 each time, analogous to the cyclops experiment proposed by Hoult (8).

As an example, in Fig.2.17 the quadrature Jeener spectrum of the tricyclodecane derivative sketched in the inset is shown. Some couplings, for instance those between protons A and B, A and K , and B and K are clearly vi-

Table 2.2 Phases ϕ_1, and ϕ_2 of the radiofrequency pulses in Fig.2.14 and phase Ψ of the receiver, through which the experiment has to be cycled for detecting the coherence transfer echo selectively and suppression of axial peaks.

Experiment number	ϕ_1	ϕ_2	Ψ
1	x	x	x
2	x	y	-x
3	x	-x	x
4	x	-y	-x

sible. Cross peaks close to the diagonal are obscured by the absolute value tailing of the diagonal peaks. As discussed in the previous section, all diagonal peaks, originating from parallel transitions, have identical phases. Therefore the absolute value calculation will show strong tailing for these diagonal resonances. The cross peaks are naturally grouped into two-dimensional spin multiplets, in the simple case of Fig.2.15 a square pattern centered at (δ_A,δ_X) or at (δ_X,δ_A), with a splitting J in both dimensions. The intensities of the individual components alternate as indicated in Fig.2.15, with the consequence that at large distances from these multiplet centres the dispersion contributions will cancel each other. Therefore these cross multiplets will not show pronounced tailing in the absolute value mode. Another, more important consequence, is that if the absorption mode of two antiphase multiplet components overlaps, partial cancellation will occur, and in the limit the cross multiplet will vanish. It is therefore essential to ensure that the digitization of the multiplet is sufficiently fine in both dimensions that mutual cancellation is avoided. This condition sets limits to the minimum sampling times needed along both time axes. In practice the length of the sampling times should at least be of the order J^{-1}, in order to be able to observe a cross peak originating from nuclei with coupling J.

When a wide range of chemical shifts has to be covered, this entails a large data matrix, which is one of the principal limitations on the method. As shown below, the antiphase nature of the multiplet components may be put to good use in order to emphasize the cross peaks in comparison with diagonal peaks. Consider again a two-spin system AX with scalar coupling J. The transition frequencies of nucleus A are Ω_{12} and Ω_{34}, the frequencies of nucleus X are Ω_{13} and Ω_{24}. Four signals contribute to the AX cross multiplet: $s_{12,13}(t_1,t_2)$, $s_{12,24}(t_1,t_2)$, $s_{34,24}(t_1,t_2)$ and $s_{34,13}(t_1,t_2)$. Because the

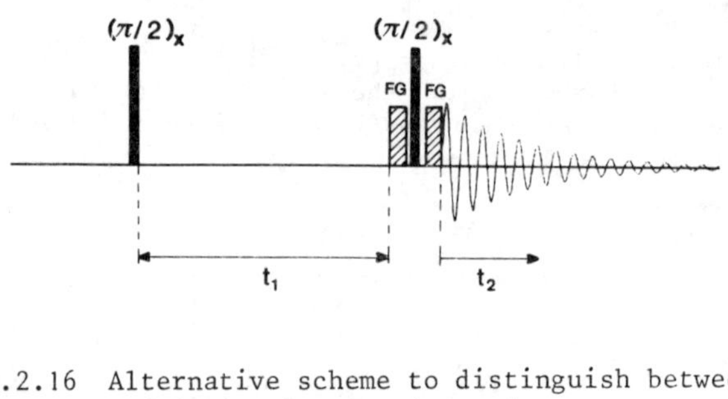

Fig.2.16 Alternative scheme to distinguish between positive and negative modulation frequencies and to suppress axial peaks using pulsed field gradients (FG). The two pulsed gradients of equal length, intensity and polarity select the coherence transfer echo component.

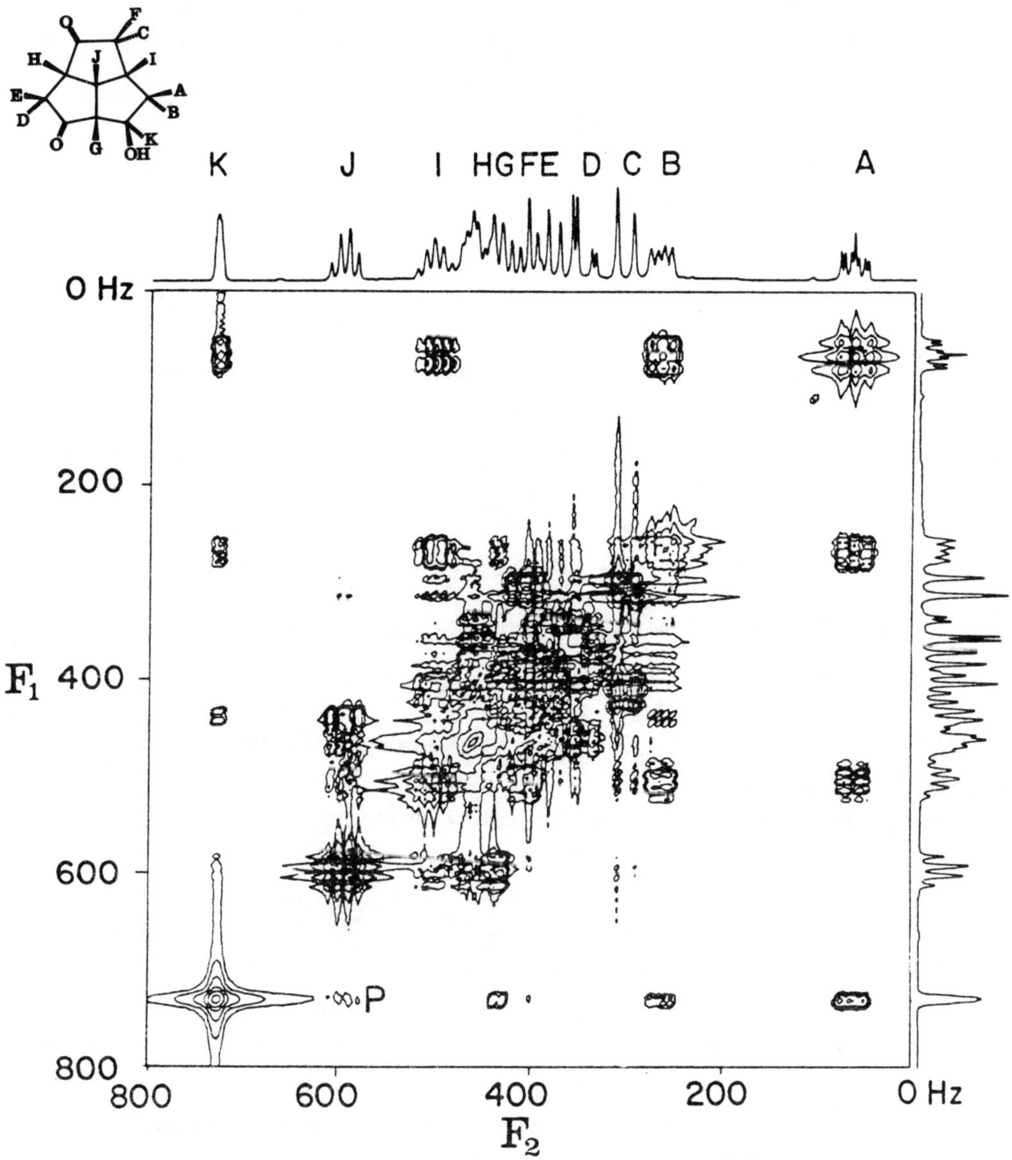

Fig.2.17 Quadrature Jeener spectrum of the tricyclodecane derivative shown on the inset, presented as an intensity contour plot.
At the bottom left, marked P, is a spurious peak generated by the overlap of the tails of two diagonal peaks. Gaussian weighting has been used to avoid truncation effects. The spectrum is recorded at 200 MHz. The sample was kindly provided by Mr. P.L. Beckwith, Dyson Perrins Laboratory, Oxford. From ref.(17).

first and the third signal originate from regressively connected transitions the phases ϕ and Ψ in Eq. [2.14] equal -90°. The second and fourth element originate from progressively connected transitions and in these cases ϕ and Ψ equal 90°. Substitution of $\Omega_{12} = \Omega_A + \pi J$, $\Omega_{34} = \Omega_A - \pi J$, $\Omega_{24} = \Omega_B - \pi J$ and $\Omega_{13} = \Omega_B + \pi J$ in Eq. [2.14] and adding the four signals together gives for the total signal contributing to the cross multiplet

$$s(t_1,t_2) = C \sin(\pi J t_1)\sin(\pi J t_2)\exp(-i\Omega_A t_1)\exp(i\Omega_X t_2) \times \times \exp(-t_1/T_2)\exp(-t_2/T_2), \quad [2.15]$$

where C is a constant.

In this equation also decay due to relaxation with time constant T_2 is taken into account.

Four signals contributing to the diagonal A multiplet are : $s_{12,12}(t_1,t_2)$, $s_{12,34}(t_1,t_2)$, $s_{34,34}(t_1,t_2)$ and $s_{34,12}(t_1,t_2)$. They all originate from parallel transitions and will have the value zero for the phase constants ϕ and Ψ. The sum of these four signals is then described by

$$s(t_1,t_2) = C' \cos(\pi J t_1)\cos(\pi J t_2)\exp(-i\Omega_A t_1)\exp(i\Omega_X t_2) \times \times \exp(-t_1/T_2)\exp(-t_2/T_2), \quad [2.16]$$

where C' is a constant with the same magnitude but different phase as the constant C in Eq. [2.15].

From Eqs. [2.15] and [2.16] it then follows that time domain weighting functions which increase with time and reach a maximum after an interval of the order of $(2J)^{-1}$ favour cross peaks at the expense of diagonal peaks. Two filtering functions, designed to achieve this are the convolution difference filter (25) and the pseudo-echo filter (26) (section 1.7). In the convolution difference filter, possibly used in combination with exponential or Gaussian weighting, the time constant T_{CD} as defined in Eq. [1.78] is chosen to be equal to $(2J)^{-1}$, where J is the largest coupling constant in the molecule. Since the envelope of the time domain data is close to that of a pseudo-echo (section 1.7.3), the tailing of the absolute value mode representation of the diagonal peaks in the correlation spectrum is partially suppressed. Since the shape of the filter is close to that of the envelope of the transferred signal, the weighting can be close to matched filtering and a gain in sensitivity for the informative cross peaks is possible.

A window which modifies the time domain signal to a Gaussian pseudo-echo (section 1.7.3) gives pure absorption line shapes in the absolute value mode, removing the tailing completely. As a demonstration, the data which resulted in the spectrum of Fig.2.17 have been reprocessed, using the Gaussian pseudo-echo window, yielding the spectrum of Fig.2.18. From this spectrum the presence of the couplings presented in the left column of Table 2.3 was determined. Since the time domain signal is symmetric with respect to t_1 and t_2, identical acquisition times and digital filtering along both time axes will usually be chosen, leading to a spectrum with exact symmetry about the diagonal (see Fig.2.18).

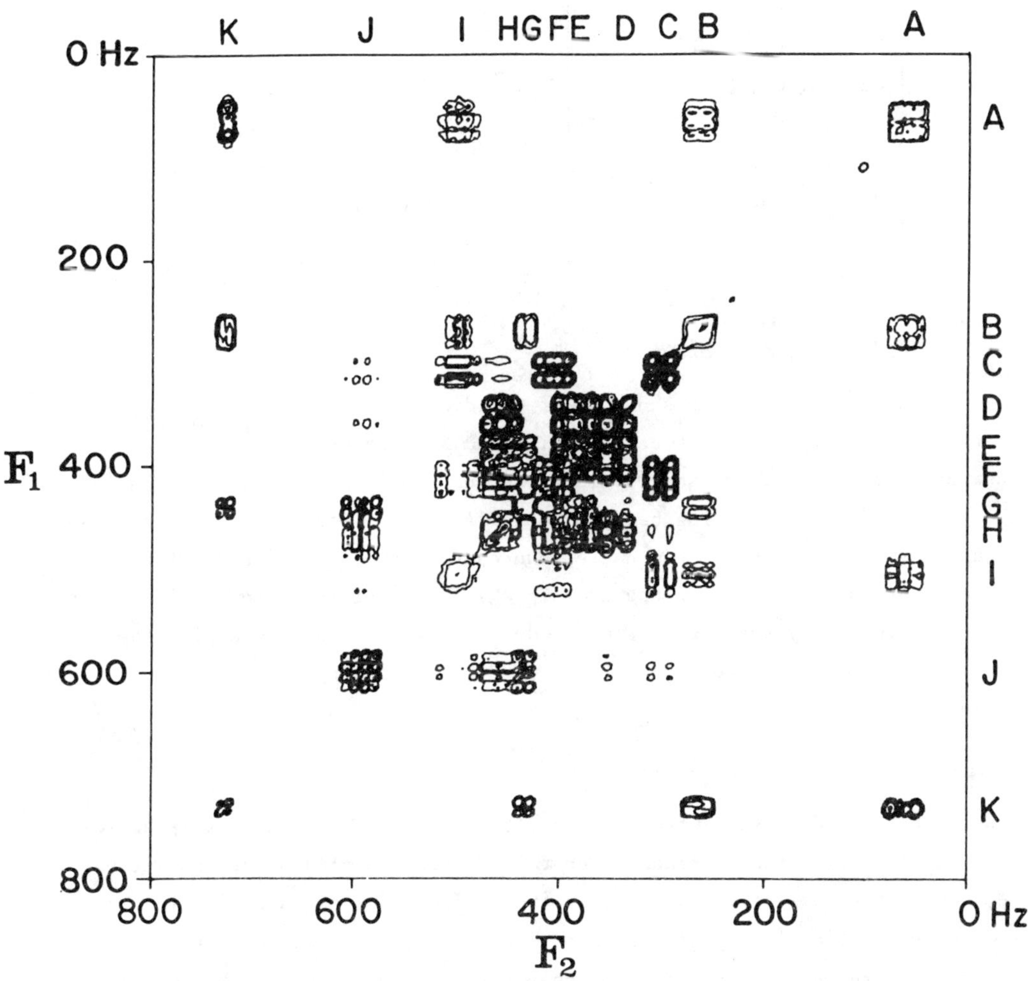

Fig.2.18 Quadrature Jeener spectrum of the tricyclodecane derivative shown as a contour plot. The pseudo Gaussian echo filter (Eq.[1.77b]) was used in both time dimensions. The cross peaks indicate the spin couplings listed in the left-hand column of Table 2.3. From ref.(17).

Table 2.3 Cross peaks indicating couplings as observed between protons in the tricyclodecane derivative sketched in the inset in Fig.2.17. The left column shows the cross peaks as found in Fig.2.18 while the right coulumn shows the extra cross peaks visible in Fig.2.23.

Normal couplings	Very weak couplings
AB, AI, AK	AC, AJ
BA, BG, BI, BK	BJ
CF, CH, CI, CJ	CA
DE, DH, DJ	DG
ED, EG, EH	
FC, FH, FI	FK
GB, GE, GJ, GK	GD
HC, HD, HE, HF, HJ	
IA, IB, IC, IF, IJ	IK
JC, JD, JG, JH, JI	JA, JB, JK
KA, KB, KG	KF, KI, KJ

Interpretation of the spectrum

Note that if a cross peak shows up at coordinates (δ_A, δ_X) and (δ_X, δ_A) that coupling is present between nuclei A and X. However, if no cross peaks show up this does not necessarily mean that no coupling is present. Even in the case of a large coupling between nuclei A and X, short transverse relaxation times of one of these two nuclei can cause such a rapid decay of the signal that the intensity of the cross peaks is surprisingly low. This effect can be increased by the use of the pseudo-echo window. As an example, the cross peak between nuclei J and I in Fig.2.18 has a very low intensity, while the coupling constant is about 9 Hz.

2.3.3 Flip angle effects

When the second pulse of the sequence (the mixing pulse) has a flip angle $\pi/2$ then magnetization is transferred to all other coupled spins, but it has been shown theoretically (15) that if the mixing pulse has a small flip angle then magnetization is transferred predominantly to directly connected transitions, those which share a common energy level. This considerably simplifies the resulting two-dimensional spectrum. Furthermore, the transferred magnetization is distributed over a smaller number of cross peaks and sensitivity can be improved, provided that the flip angle is not too small. In practice a $\pi/4$ pulse is a good compromise, giving both simplification and good sensitivity.

This flip angle effect can be visualized by considering the mixing pulse to be decomposed into a cascade (27) of selective pulses, each affecting only a single transition. Consider the three-spin AMX system with energy level diagram shown in Fig.2.19. The initial $\pi/2$ pulse is not decomposed into a cascade, but simply creates precessing transverse magnetization in all the allowed transitions.

The mixing pulse of flip angle α applied to the AMX spin system of Fig. 2.19 may be represented as a cascade of twelve selective pulses:

$$\alpha(12)\ \alpha(46)\ \alpha(35)\ \alpha(78)$$
$$\alpha(25)\ \alpha(13)\ \alpha(47)\ \alpha(68)$$
$$\alpha(26)\ \alpha(58)\ \alpha(14)\ \alpha(37)$$

Each of these selective pulses can be described by an exponential fictitious spin-½ operator (28,29,30, App.III). For the present purpose, however, this is not necessary. Using the arithmetic of fictitious spin-½ operators it can be derived that all the pulses affecting the A spins must be grouped together, and similarly the pulses affecting the M spins and the X spins form separate groups (App.III). However, the order of the three groups and the order within any group can be permuted.

Attention may be focussed on the fate of one typical transverse magnetization, $A_{tr}(12)$, and although this precesses as a function of t_1, its amplitude R_o is all the needs to be considered here. Consider, first of all, magnetization transfer to a typical connected transition $M_{tr}(25)$. Not all the elements in the pulse cascade are involved, and it is sufficient to calculate

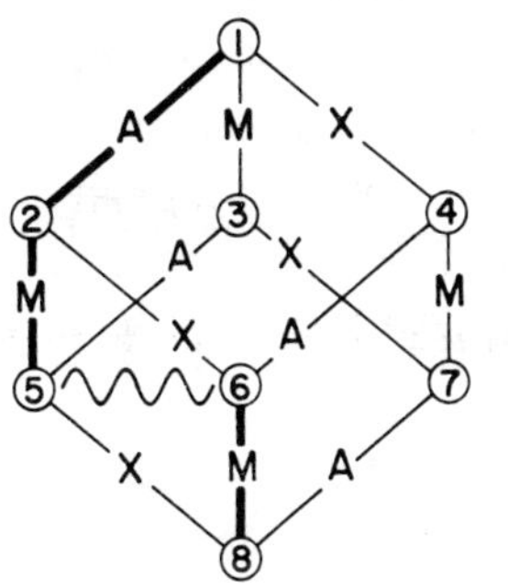

Fig.2.19 The energy level diagram of a proton AMX system. Transfer of magnetization from A(12) to M(25) and M(68) is discussed in the text. From ref.(17).

the effect of the cascade:

$$\alpha(12)\ \alpha(25)\ \alpha(26)\ \alpha(58).$$

The effect of a transition-selective pulse on a magnetization component is explained in Appendix III. Suppose, for the purpose of convenience that the α pulse is applied along an axis perpendicular to the orientation of transverse magnetization component $A_{tr}(12)$. The first element $\alpha(12)$ converts transverse magnetization $A_{tr}(12)$ into longitudinal magnetization $A_z(12)$ of amplitude $R_o \sin\alpha$ (App.III). The second element $\alpha(25)$ converts one half of this into transverse magnetization $M_{tr}(25)$ of amplitude $\frac{1}{2}R_o\sin^2\alpha$. The remaining two pulses withdraw magnetization from $M_{tr}(25)$ (creating multiple quantum coherence) leaving an amplitude $\frac{1}{2}R_o\sin^2\alpha\cos^2(\alpha/2)$ (App.III).

A different sequence of events governs the transfer to a non-connected transition $M_{tr}(68)$, although the same four elements of the pulse cascade are applicable. Transverse magnetization $A_{tr}(12)$ is converted into $M_{tr}(25)$ by the first two pulses with amplitude $\frac{1}{2}R_o\sin^2\alpha$. But the third element $\alpha(26)$ creates zero quantum coherence ZQ(56) with an amplitude $\frac{1}{2}R_o\sin^2\alpha\sin(\alpha/2)$ (App.III) and the last element $\alpha(58)$ reconverts this into observable transverse magnetization $M_{tr}(68)$ with an amplitude $\frac{1}{2}R_o\sin^2\alpha\sin^2(\alpha/2)$. A reversal of the sense of the last two elements of the cascade would have given the same result but by way of the double quantum coherence DQ(28). It is a general rule that transfer to non-connected transitions involves double or zero quantum coherence as an intermediate, whereas directly connected transitions acquire magnetization by conversion of the longitudinal magnetization of a connected transition.

As a result, the ratio of the intensities of connected cross peaks to that of non-connected cross peaks is given by $\cot^2(\alpha/2)$. Thus for a mixing pulse of 45° the connected transitions are 5.8 times stronger. This is the setting used in many of the experiments described below when good discrimination between connected and non-connected transitions is required, as in the determination of relative signs of coupling constants. When sensitivity is an important consideration, a mixing pulse of 60° delivers higher intensities in the cross peaks of the directly connected transitions. The intensities of the indirectly connected transitions are in this case still a factor of 3 lower.

Similar pulse cascade arguments can be used to predict the relative intensities of parallel transitions, where magnetization is transferred between component lines of the same spin multiplet. These lines are situated just off the principal diagonal within a distance of the order of J. The general rule

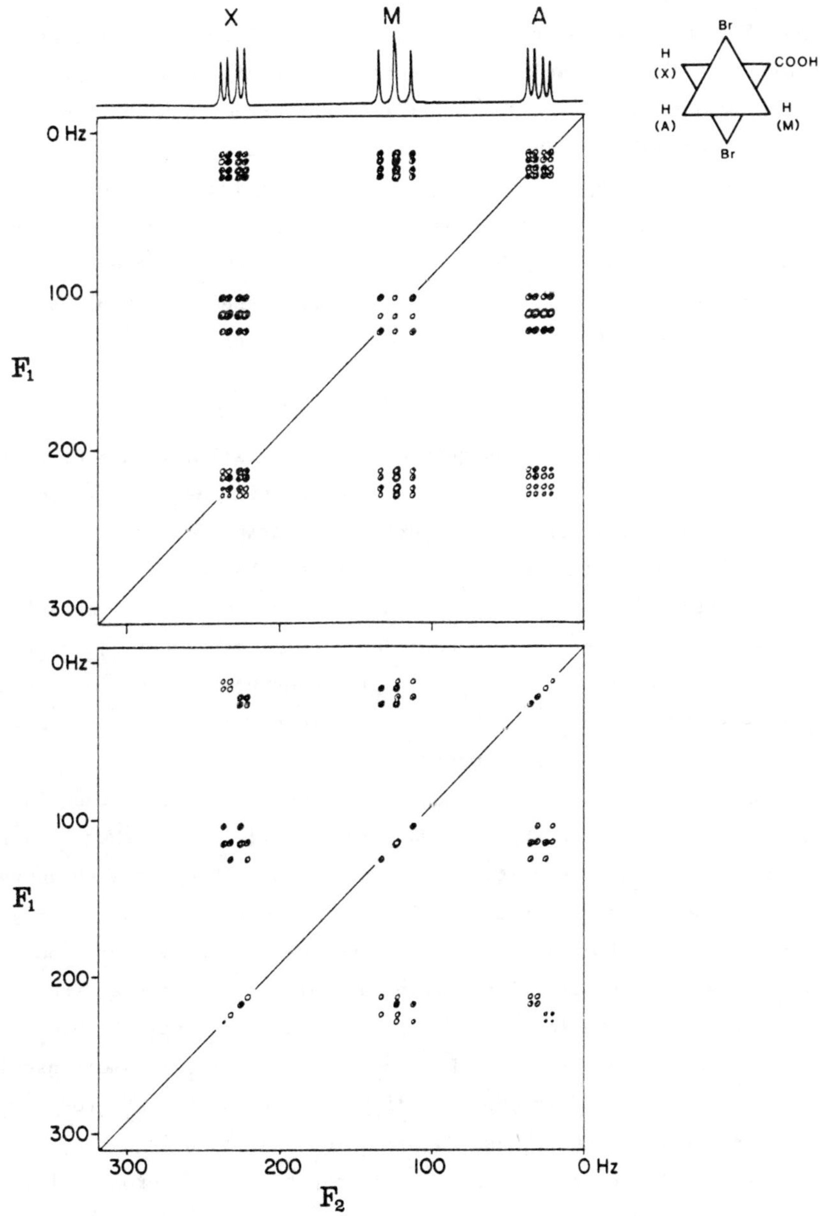

Fig.2.20 Contour plots of the Jeener spectra of 2,3-dibromopropionic acid, an AMX spin system. The upper spectrum was obtained with a $\pi/2$ mixing pulse. The lower spectrum was obtained with a $\pi/4$ mixing pulse, so that only directly connected transitions are intense enough to show contours. (Spectra recorded at 200 MHz). From ref.(17).

is that parallel transitions lose intensity compared with connected transitions as the flip angle is reduced below $\pi/2$, but that certain parallel transitions lose intensity faster than others, because they involve more stages of magnetization transfer in succession. For example the transfer from A(12) to A(34) gives A(34) magnetization equal to $R_o \sin^2(\alpha/2)\cos^2(\alpha/2)$, while an amount equal to $R_o \sin^4(\alpha/2)$ is transferred to A(78). In practice it can be very useful to reduce the relative intensity of parallel transitions so as to leave the region close to the principal diagonal clear in order to search for cross peaks. Fig.2.20 shows an example of a spectrum where most of the parallel transitions have been reduced below the threshold of the lowest intensity contour. These results based on the concept of pulse cascades are consistent with the results derived more formally by Aue et al.(15).

2.3.4 Relative signs of coupling constants

There is an interesting consequence of using small flip angles to restrict the transfer of magnetization to connected transitions. In a system of three or more coupled spins, if the appropriate splittings can be resolved, the relative signs of coupling constants may be ascertained by inspection of the Jeener spectrum. The basis of the method is the same as that of selective decoupling experiments (31,32). Consider the AMX system of 2,3 dibromopropionic acid, where conventional double resonance and double quantum experiments have established that the two vicinal couplings have opposite signs to that of the geminal coupling (33,34). When the flip angle of the mixing pulse is small, then AX cross peaks involving a simultaneous flip of the M nucleus are of vanishingly small intensity; on the other hand for a flip angle of $\pi/2$ all 16 multiplet components are observed with comparable intensity. The Jeener spectrum obtained with a small flip angle (Fig.2.20) thus resembles a selective double resonance experiment where A and X are decoupled only in molecules with one of the spin states of the M nucleus, say M(α) but not M(β). The local magnetic fields at the sites of the A and X nuclei due to the couplings to the M spin must be in the same sense if $J_{AM} \cdot J_{MX} > 0$. Thus the corresponding displacements of the AX cross peak are in the same sense in the F_1 and F_2 dimensions. As a consequence, the line connecting the two centres of gravity of the two AX sub-cross multiplets, makes an angle smaller than 45° with the principal diagonal of the spectrum. On the other hand, if J_{AM} and J_{MX} have opposite signs, the AX multiplets are tilted in the opposite sense, and the line joining the centres of gravity makes an angle greater than 45° with the principal diagonal. If either J_{AM}

or J_{MX} is vanishingly small, then the AX multiplet lies vertically on one side of the diagram and horizontally on the other, and no sign determination is possible.

Fig.2.20 shows the contour diagram for 2,3 dibromopropionic acid, first for a mixing pulse flip angle of $\pi/2$, showing each cross peak made up of 16 components[★], then for a flip angle of $\pi/4$, where each cross peak has been reduced to 8 components. (In the spectrum obtained at low flip angle, the parallel transitions - where magnetization has been transferred within a single spin multiplet - are absent, leaving only diagonal peaks which fall exactly on the principal diagonal.) Each cross peak is made up of two groups of four lines which form an exact square (compare Fig.2.15). It is the relative position of these two squares which determines the relative signs of the coupling constant, each cross peak relating the signs of two coupling constants. In the case of 2,3 dibromopropionic acid, J_{AX} and J_{MX} have like signs, opposite to the sign of J_{AM}, the geminal coupling.

Once the principle of relative sign determination has been established in a text book case like 2,3 dibromopropionic acid, it is readily applied to more complex spectra simply by noting the tilt of the pattern of lines in a cross peak. Fig.2.21 shows a contour plot of the spectrum of the tricyclodecane derivative (Fig.2.17) obtained with a mixing pulse of flip angle 45°. The tilt of several cross peaks is indicated by the arrows.Arrows making an angle smaller than 45° with the diagonal indicate like signs, those with an angle larger than 45° opposite signs; an angle of 45° means that one of the two couplings involved is vanishingly small. The relative signs determined from the spectrum in Fig.2.21 are set out in Table 2.4.

Table 2.4 Relative signs of certain coupling constants in the tricyclodecane derivative determined by inspection of the slope of cross peaks in the Jeener spectrum in Fig.2.21.

Cross peak	Couplings	Sign
AK	J(AB) and J(BK)	opposite
AI	J(AB) and J(BI)	opposite
AB	J(AI) and J(BI)	like
BK	J(AB) and J(AK)	opposite
BI	J(AB) and J(AI)	opposite
CI	J(CF) and J(FI)	opposite
DE	J(DH) and J(EH)	like

★ Due to overlap multiplets in which the M spin is involved show only 12 components in the cross multiplets and 9 in the diagonal multiplet.

The simplicity of the Jeener spectrum obtained with small flip angles for the mixing pulse stems from the fact that each cross peak is basically from an AX or AB spin system; all other splittings are passive and simply create several independent AX or AB subspectra. In a more general case, groups of two or three spins could be equivalent, leading to A_2X, AA'X spectra for example.

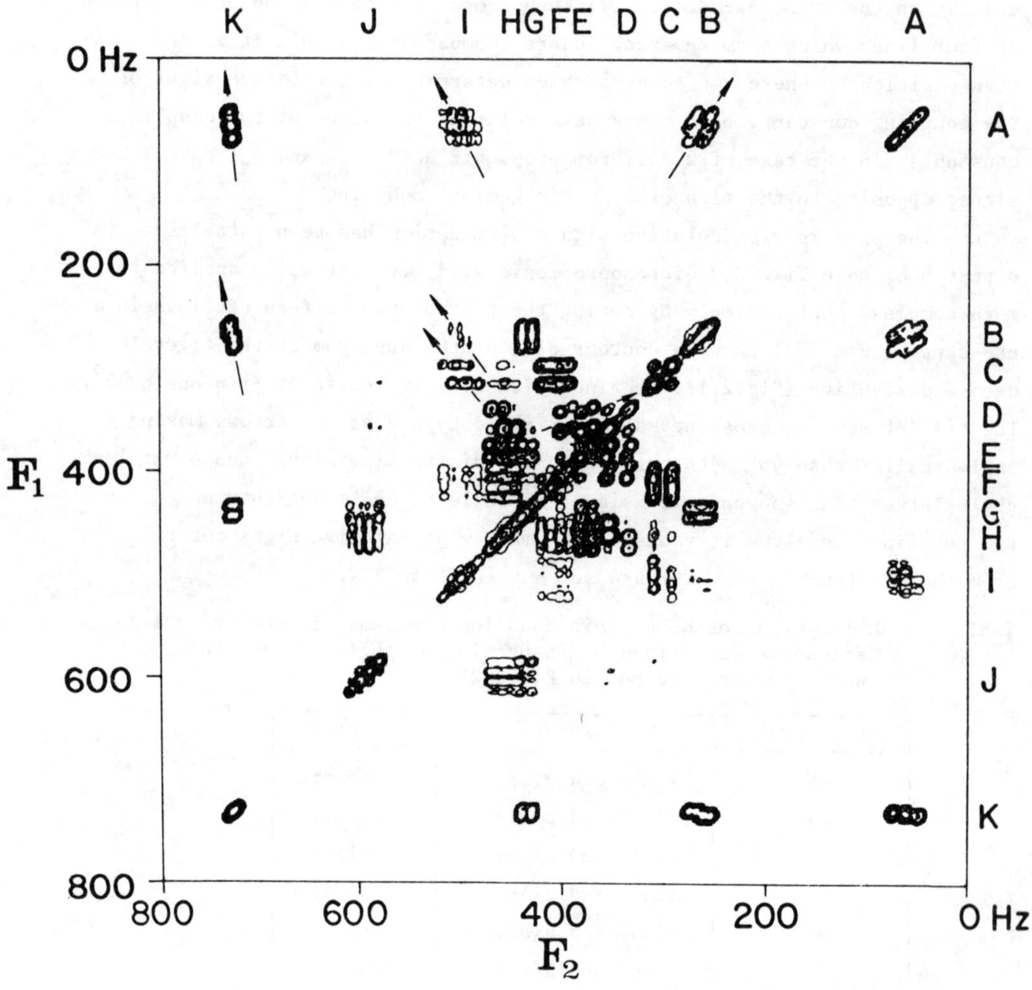

Fig.2.21 Jeener spectrum of the tricyclodecane derivative with the conditions adjusted to determine the relative signs of certain coupling constants. The mixing pulse was reduced to π/4 and a pseudo Gaussian echo filter was used in both time dimensions. The arrows indicate the approximate tilt of the cross peaks. The relative signs determined from this spectrum are set out in Table 2.4. From ref.(17).

2.3.5 Long-range coupling

As follows from the discussions in sections 2.3.1 and 2.3.2, cross peaks indicating weak long-range couplings will generally be (vanishingly) small. Not only is the transfer from A to X proportional to $\sin(\pi J_{AX} t_1)\exp(-t_1/T_2)$, which will be small for unresolved long-range couplings, but also the transferred components start out in antiphase, giving a net magnetization proportional to $\sin(\pi J_{AX} t_2)\exp(-t_2/T_2)$, so that the detected transfer will be small (Eq. [2.15]).

Assuming that $J_{AX} \ll (T_2)^{-1}$, it is easily shown that the transfer process is most efficient at time $t_1 = T_2$. Similarly it is found that the detection process is most efficient at time $t_2 = T_2$. Therefore, in the pulse scheme of Fig.2.14, using the pseudo echo window, the optimum setting for the centre of the pseudo echo is near the points $t_1 = T_2$ and $t_2 = T_2$. However, this would imply acquisition times along the t_1-axis and t_2-axis comparable to $2T_2$, leading to an unacceptably large data matrix. Fixed delays Δ can be inserted at the end of the evolution and beginning of the detection periods (Fig.2.22), changing the optimum setting for the centre of the pseudo echo to the points $t_1 = T_2 - \Delta$ and $t_2 = T_2 - \Delta$ and thus limiting the acquisition times required to $2T_2 - 2\Delta$. Note that due to the coherence transfer echo effect, the transfer and detection process is not influenced by static magnetic field inhomogeneity if diffusion in the sample is neglected.

The tricyclodecane derivative sketched in Fig.2.17 provides a rich field for the investigation of long-range couplings. An estimate of the average value of the real T_2 being about 0.45 sec., the length of the delay Δ was chosen to be equal to 0.3 sec. with acquisition times both equal to 0.3 sec. Since sensitivity is a critical problem in observing long-range couplings, the flip angle of the mixing pulse was set to be equal to $\pi/3$. Fig.2.23 shows a spectrum obtained under these conditions. The couplings which can be

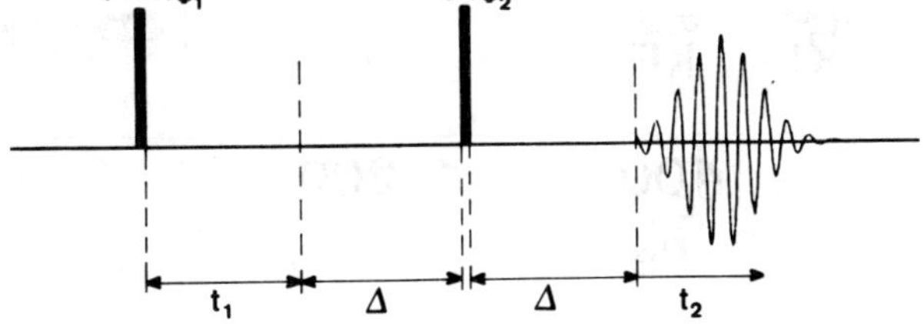

Fig.2.22 The inserting of delays with fixed length Δ in the basic pulse scheme to optimize detection of long-range couplings.

detected in this sepctrum and not in the spectrum of Fig.2.18 are set out in Table 2.3 in the column "very weak couplings". Note that cross peaks due to large couplings can be absent in this spectrum (e.g. BG) due to the fact that Jt_1 and Jt_2 are close to an integer number for $t_1, t_2 = T_2 - \Delta$, and the transfer will be small.

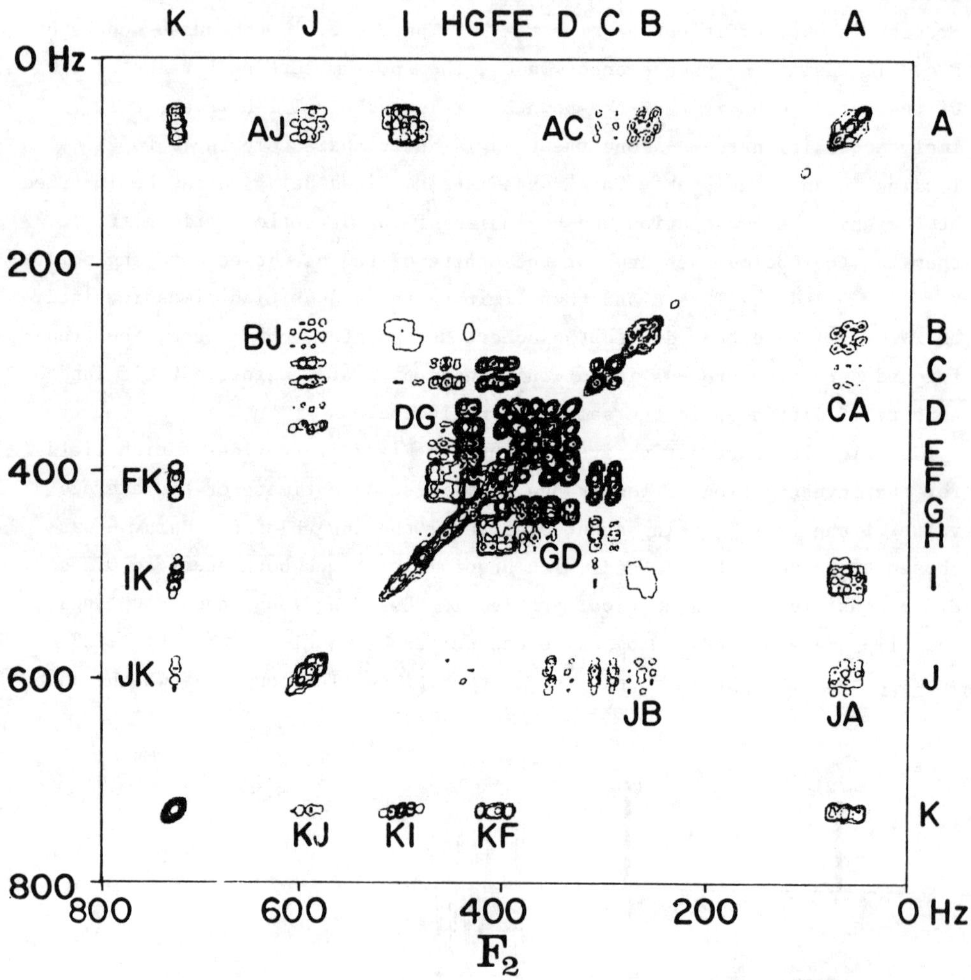

Fig.2.23 Jeener spectrum of the tricyclodecane derivative obtained under conditions calculated to emphasize long-range couplings (see text), shown as a contour plot. Pseudo Gaussian echo shaping of the responses was used in both time dimensions. From ref.(17).

2.3.6 Broad-band decoupling in the F_1 dimension

Throughout the interval between radiofrequency pulses nuclear magnetization vectors corresponding to the components of a given spin multiplet diverge continuously at a rate determined by J. Suppose that this interval were to be fixed at a value t_d seconds, and a π refocussing pulse introduced after a vaiable delay $\frac{1}{2}t_1$ sec. ($t_1 \leq 2t_d$) after the initial $\pi/2$ pulse (Fig.2.24). As will be explained in section 3.1, in the case of a homonuclear coupled spin system the angle between magnetization multiplet components in the transverse plane remains unchanged if a π pulse is applied. Then, although the amplitude of the transferred magnetization would depend on Jt_d, there would be no J modulation of the signal as t_1 was varied. On the other hand, the effect of the chemical shift would change, going from one maximum at $t_1 = 0$ to zero when $t_1 = \frac{1}{2}t_d$ (exact refocussing) and back to a maximum again at $t_1 = t_d$. This is explained in detail in ref.(35).

This is the basis of a method of broad-band decoupling in the F_1 dimension (36). It relies on the assumption that the coupling is first-order, otherwise spurious responses are excited; in an AB spin system, for example, there is a spurious response at the mean chemical shift frequency (37). By collapsing all spin multiplet structure in the F_1 dimension onto the appropriate chemical shift frequency, this greatly simplifies the Jeener spectrum, concentrating the intensity of the multiplet components. The information that is lost in that process is that discussed in the previous section - the relative signs of the coupling constants, and certain cross peaks due to large couplings for which Jt_d is an integer.

In this method it is important to employ a mixing pulse of a small flip angle ($\pi/4$) to avoid mutual cancellation of antiphase components within a given cross peak. The usual phase cycling of the $\pi/4$ pulse and the receiver reference phase (Table 2.2) ensure that the coherence transfer echo is detec-

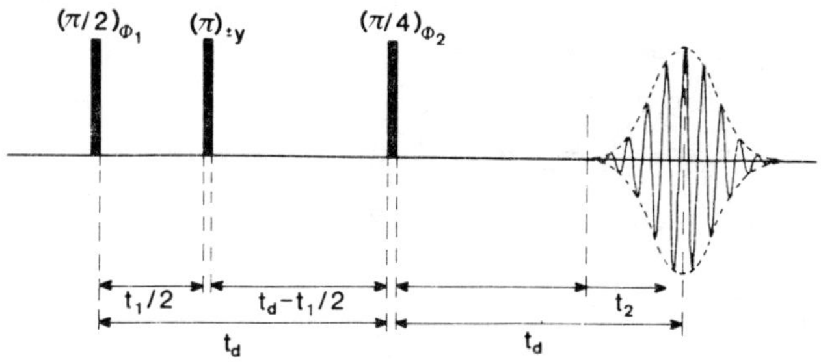

Fig.2.24 The pulse sequence used to obtain Jeener spectra that are decoupled in the F_1 dimension. The delay t_d is fixed and only t_1 and t_2 are varied. A pseudo Gaussian echo shaping function is indicated in the t_2 dimension. The phases of the $\pi/2$ pulse (ϕ_1), the $\pi/4$ pulse (ϕ_2) and the receiver (Ψ) are cycled according to Table 2.2.

ted. In addition, the π pulse is carefully calibrated and phase alternated along the ±y-axis of the rotating frame in order to avoid spurious responses due to pulse imperfections (38). Figure 2.25 shows the effect of this broadband decoupling scheme on the Jeener spectrum of the tricyclodecane derivative.

The technique provides an attractive alternative to the established method of obtaining decoupled proton spectra by 45° projection of two-dimensional J-spectra (36). The Jeener spectrum with F_1 decoupling is simply projected onto the F_1-axis. In this application highest resolution along the F_1-axis is wanted, and the sampling time along the t_1-axis (= $2t_d$) is usually chosen to be longer than the sampling time along the t_2-axis. Figure 2.25 shows a projection of the two-dimensional spectrum onto the F_1-axis together with an indication of the spurious lines arising from strong coupling between protons D and E. Since the π pulse reverses the angle through which a magnetization component has precessed just before the π pulse, the sense of the principal diagonal in the spectrum is also reversed.

A cross section through this Jeener spectrum at an appropriate point provides a clear indication of the pattern of couplings to a given proton. For example, if it is required to know which protons are coupled to proton J, a single vertical cross section is taken so as to cut the principal diagonal at the chemical shift of J. In practice it is rather better to extract a vertical band of frequencies from the two-dimensional matrix (consisting of a number of cross sections), as indicated by the dashed lines in Fig.2.25; the band is then projected horizontally onto the F_1-axis. This has the effect of summing all the individual component lines in a given cross peak. The result is a one-dimensional spectrum (Fig.2.26) which indicates the cross peaks associated with proton J, each cross peak appearing now as a single line. Not only does this show that proton J is coupled to A, B, C, D, H , I and K, but it also gives an approximate estimate of the relative magnitudes of the long-range couplings, as the acquisition parameters were optimized for couplings of the order of the natural line width. The intensity I of a peak depends analogous to Eq. [2.15] on the coupling J according to

$$I = C \sin^2(Jt_d)\exp(-2t_d/T_2). \qquad [2.17]$$

The problem in using Eq. [2.17] for an exact determination of the magnitude of the coupling is that the constant C depends on the number of other nuclei coupled to the nuclei under investigation and on the magnitudes of these other couplings.

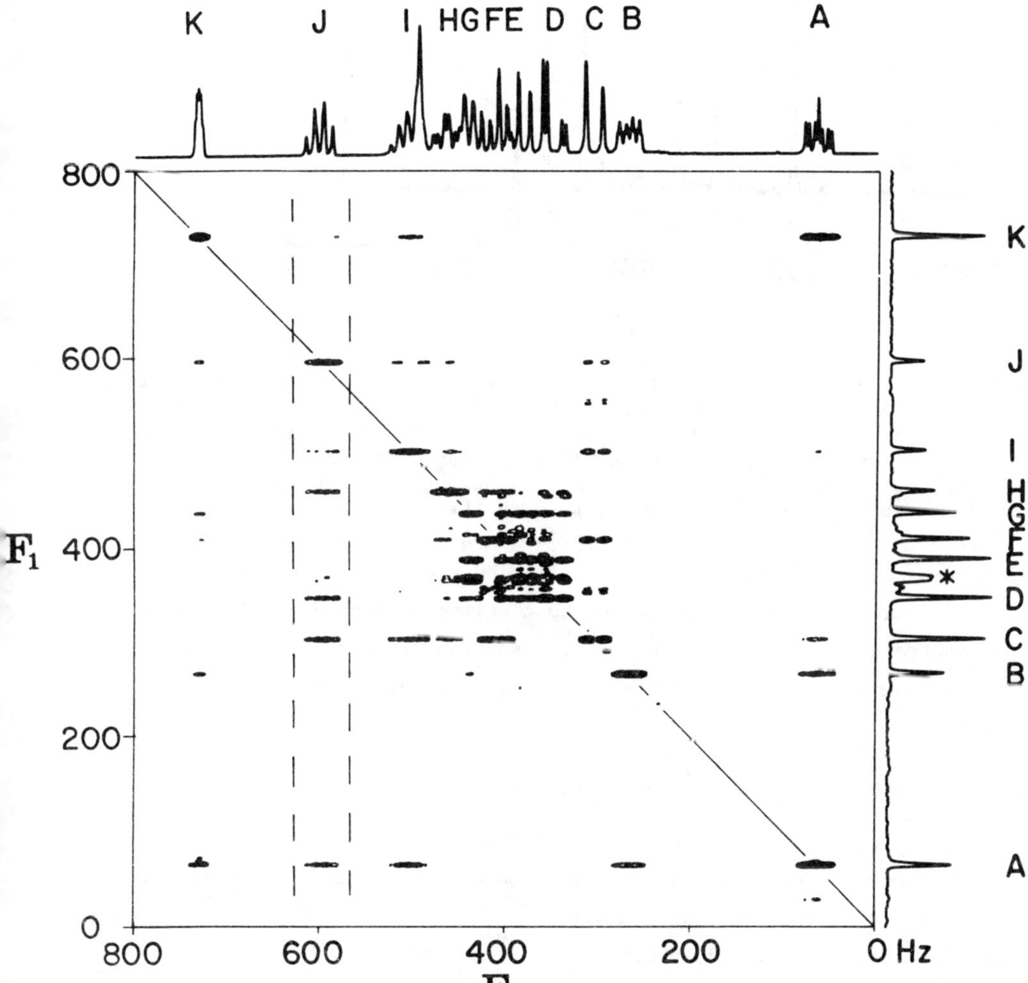

Fig.2.25 Jeener spectrum of the tricyclodecane derivative showing broadband decoupling in the F_1 dimension through the use of the pulse sequence shown in Fig.2.24. The spectrum running along the top of the diagram is the fully-coupled spectrum; that shown in the right margin is the projection of the two-dimensional spectrum. The overlapping lines in this projection which are due to strong coupling between protons D and E are marked with an asterisk. A band of signals associated with proton J has been picked out of the data matrix and projected onto the F_1-axis in order to obtain the spectrum shown in Fig.2.26. From ref.(17).

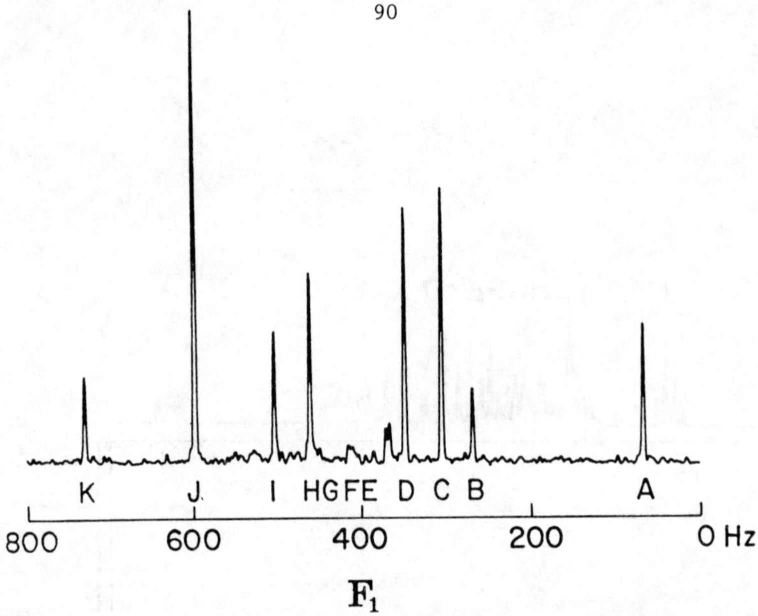

Fig.2.26 The projection onto the F_1-axis of a narrow band of signals extracted from the data matrix of Fig.2.25 representing all the cross peaks associated with proton J. This indicates that J is coupled to protons A, B, C, D, H, I and K, and some indication of the relastrengths of these couplings can be obtained provided that they are weak. An artefact appears between the chemical shift frequencies of protons D and E because these protons are strongly coupled.

In agreement with Eq.[2.17] large coupling constants give anomalous intensities, for example the cross peak due to coupling between J and G is absent in this trace. On the other hand, a weak spurious response is observed at the mean chemical shift of protons D and E, attributable to strong coupling effects.

2.3.7 The Secsy and Focsy experiments

Two modifications on Jeener's original experiment have been proposed by Ernst and co-workers. Both modifications have the purpose of limiting the size of the data storage space required. The first modification proposed was named Spin echo correlated spectroscopy (Secsy) (18,20). More recently another modification has been proposed, called Foldover corrected correlated spectroscopy (Focsy) (20). Both modifications employ basically the same two-pulse scheme as initially proposed by Jeener (14) (Fig.2.14). In both the Secsy and Focsy experiments the coherence transfer echo is detected selectively by cycling the phases of the second radiofrequency pulse and of the receiver in the way described in section 2.3.2 (Table 2.2). It will be shown that both modifications proposed can only be advantageous in certain uncommon cases, while in general the quadrature Jeener experiment as proposed in section 2.3.2 is a better alternative.

The Secsy experiment

The Secsy experiment employs the standard pulse sequence of Fig.2.14, but with acquisition (and thus t_2) starting at the midpoint of the spin echo, at time $t_2 = t_1$ in the original experiment (Fig.2.27). Substitution of $t_2 = t_2' + t_1'/2$ and $t_1 = t_1'/2$ in Eq. [2.14] gives for the new time domain signal

$$s_{kl}(t_1',t_2') = \sum_{mn} iC_{kl,mn}\exp\{-i(\Omega_{mn} - \Omega_{kl})t_1'/2 + i\Omega_{kl}t_2' + + i\Psi_{kl,mn} - i\phi_{kl,mn}\}. \qquad [2.18]$$

As is clear from this equation, the acquired signals are modulated in phase as a function of t_1' by frequency $(\Omega_{mn} - \Omega_{kl})/4\pi$, which is thus equal to half the difference in frequency between the two correlated transitions. Assuming identical transverse relaxation times T_2 for transition kl and mn, the signal decay as a function of t_1 is in the quadrature Jeener experiment and in the Secsy experiment described by the factors $\exp(-t_1/T_2)$ and $\exp(-t_1'/T_2)$. Neglecting the effect of field inhomogeneity, the line width along the F_1-axis in the Secsy spectrum is thus the same as in the quadrature Jeener spectrum, but the difference in frequency between the two resonances, and thus the resolution, is halved. This is a disadvantage because high resolution along the F_1-axis is necessary to avoid mutual cancellation of cross peaks which have opposite intensity. A second disadvantage of this experiment is that the first half of the coherence transfer echo is not acquired, decreasing the sensitivity significantly. Diagonal peaks as mentioned in section 2.3.1 now lie on the line $F_1 = 0$. As an illustration a Secsy correlation spectrum of 2,3 dibromopropionic acid is shown in Fig.2.28.

As follows from Eq. [2.18] the spectral width along the F_1-axis (and thus the sampling frequency $(\Delta t_1')^{-1}$ along the t_1-axis) has to be larger than the maximum difference in chemical shift frequency between coupled nuclei, in

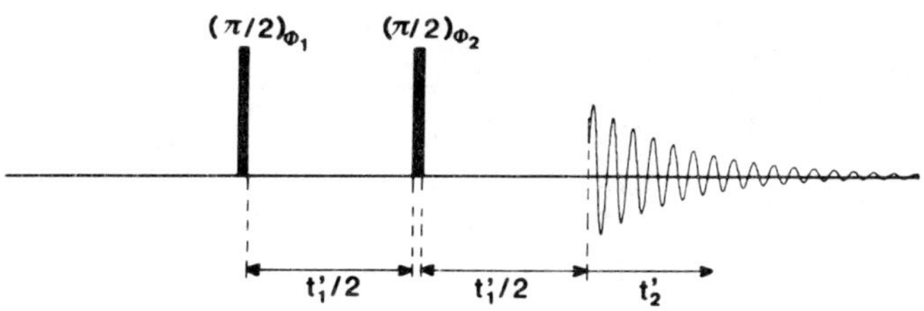

Fig.2.27 Pulse scheme of the Secsy experiment. The phases of the pulses and the phase Ψ of the receiver are cycled according to Table 2.2.

order to avoid folding. This compares with twice the maximum offset frequency of a resonance line of interest in the quadrature Jeener experiment, as discussed in section 2.3.2. Hence, the sampling frequency along the t_1-axis in the Secsy experiment can often be lower. However, since resonance lines along the F_1-axis are only separated by half their difference in transition frequency, the resolution is a factor of two lower. In order to double the resolution, a sampling time along the t_1-axis at least twice as long compared with the quadrature Jeener experiment is required.

In conclusion it can be said that in order to obtain a reduction of the size of the data matrix by using the Secsy experiment, maintaining the same

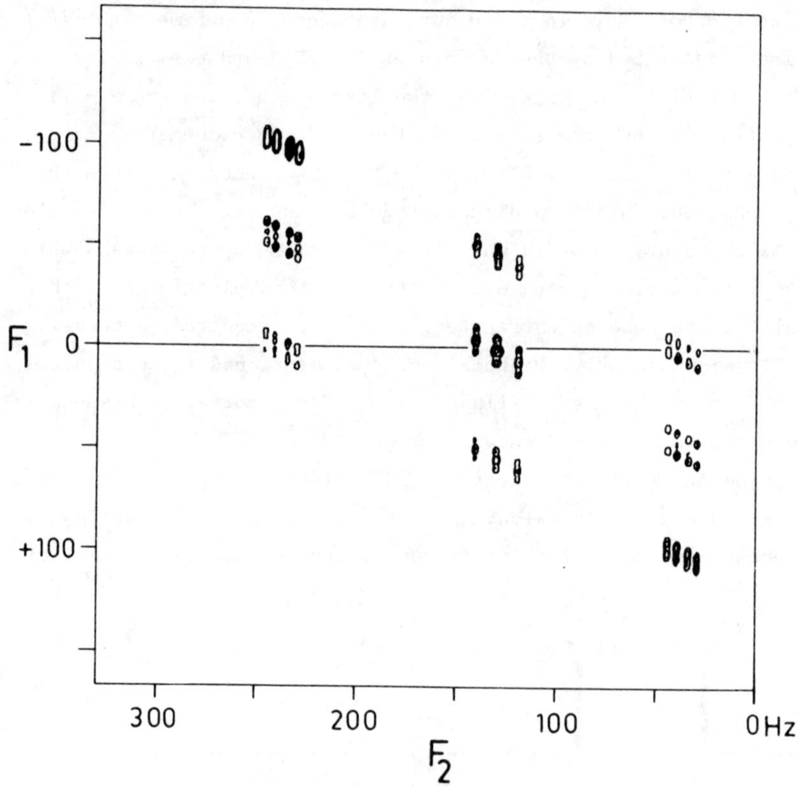

Fig.2.28 Homonuclear correlation spectrum of 2,3-dibromopropionic acid, obtained with the Secsy experiment. The diagonal in the normal Jeener spectrum now lies on the line $F_1 = 0$. A cross peak at (F_1, F_2) indicates coupling between nuclei with resonance frequencies F_2 and $2F_1 + F_2$. (Compare Fig.2.20). (Spectrum recorded by T. Mareci).

digital resolution, the maximum difference in frequency between two coupled nuclei has to be less than half the spectral width. This is not commonly the case in practice. The main disadvantages of the Secsy experiment are the loss in resolution (due to relaxation in the second half of the t_1'-period) and the loss in sensitivity due to the fact that the first half of the coherence transfer echo is not sampled.

The Focsy experiment

The Focsy experiment has been proposed more recently (20) and gives the same resolution and sensitivity as the quadrature Jeener experiment. Actually, the difference between these two experiments is only a software routine for spectrum modification , applied after a normal quadrature Jeener spectrum is obtained as described in section 2.3.2. The computer routine has the purpose of correcting folding errors along the F_1-axis. This routine changes the coordinates in the 2D frequency spectrum according to $(\omega_1', \omega_2') = (\omega_1 - \omega_2, \omega_2)$. The diagonal which is originally on the line $\omega_1 = \omega_2$, is now transferred to $\omega_1 = 0$, and the modified spectrum looks similar to a Secsy spectrum. As the authors point out, the main trick is that if $|\omega_1 - \omega_2|/2\pi$ is larger than half the sampling frequency along the t_1-axis, the ω_1'-coordinate is increased or decreased by an integer multiple of $2\pi(\Delta t_1)^{-1}$ in such a way that it fits again into the normal spectral region, and certain folding errors are corrected. A closer analysis shows that the sampling frequency along the t_1-axis in this case has to be larger than twice the maximum difference in shift frequency between two coupled nuclei. Comparing this with the sampling frequency in the quadrature Jeener experiment, which is equal to the spectral width of the region of interest, shows that a reduction of the size of the data matrix is only possible if the largest difference in shift frequency between two coupled nuclei is smaller than half the spectral width of the region of interest. Since this is not commonly the case in practice, this modification is not recommended either for general use.

The interpretation of frequencies measured in Secsy or Focsy 2D spectra is less straightforward than in the quadrature Jeener spectrum, since two transition frequencies now determine the frequency of a cross peak along the F_1-axis.

2.4 Correlation of chemical shifts through exchange and cross relaxation

Apart from the scalar interaction, signals originating from nuclei can be correlated because the nuclei exchange their position in the molecule or because of cross relaxation between the nuclei. Determination of cross relaxation effects appears to be a very powerful tool in solving assignment and conformation problems in large biochemical molecules as proteins. A scheme to detect the presence of exchange or cross relaxation non-selectively by means of a two-dimensional experiment has been developed by Ernst, Jeener et al.(39,40,41). Because nothing has been contributed to this technique by the author and because a detailed theoretical description of the experiment exists (39), only a short and simplified description of this important experiment will be given here.

Before describing the experiment, it should be mentioned that the effects of chemical exchange and cross relaxation on the eventual 2D spectrum are analogous. Cross relaxation is caused by dipole-dipole coupling and cross relaxation rates are a measure of internuclear distance and of mobility in a molecule, and do not depend on the number of bonds involved. This means that different information compared with the methods described in sections 2.2 and 2.3 is obtained.

The pulse scheme is shown in Fig.2.29. Consider a spin system consisting of two isolated homonuclear spins A and X, without any scalar interaction but with mutual cross relaxation. The effect of spin A on spin X during the pulse sequence is briefly discussed below. In the case where the first two $\pi/2$ pulses are both applied along the x-axis, the longitudinal magnetization of nucleus A, $M_{z_A}(t_1)$, just after the second $(\pi/2)_x$ pulse is given by

$$M_{z_A}(t_1) = -M_{o_A}\cos(\Omega_A t_1), \qquad [2.19]$$

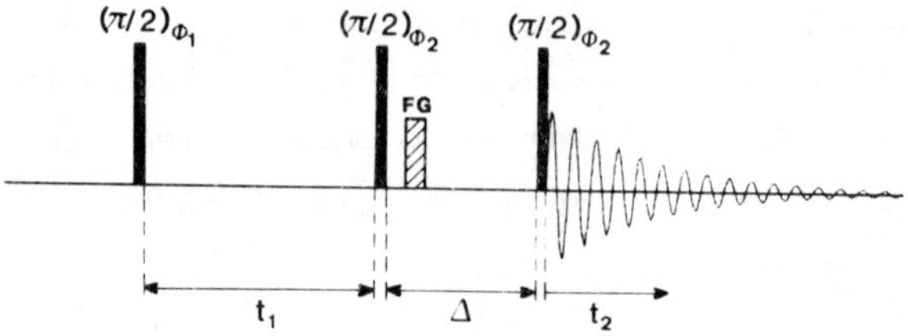

Fig.2.29 Scheme for homonuclear correlation spectroscopy based on exchange or cross-relaxation. During the delay Δ a pulsed field gradient is applied. To distinguish between positive and negative modulation frequencies and to eliminate axial peaks, the phases ϕ_1,ϕ_2 and the receiver phase Ψ are cycled according to Table 2.2 (section 2.3.2).

where M_{oA} and Ω_A are the thermal equilibrium longitudinal magnetization and the angular chemical shift frequency of nucleus A. During the mixing period with length Δ, cross relaxation with nucleus X takes place, changing the longitudinal X magnetization by an amount $C M_{zA}(t_1)$, where C is a constant depending on the cross relaxation rate. Just before the final pulse the longitudinal X-spin magnetization is thus given by

$$M'_{zX}(t_1) = f(M_{zX}(t_1)) + C M_{zA}(t_1), \qquad [2.20]$$

where $f(M_{zX}(t_1))$ is a function depending on the relaxation of nucleus X during the delay Δ and the distortion of the X-spin magnetization by the first two pulses in the sequence. It is the second term at the right-hand side of Eq. [2.20] that is the term of interest, since this is due to the cross relaxation from nucleus A to X. A third $(\pi/2)_x$ pulse converts this term into transverse X magnetization which is, using Eqs. [2.19] and [2.20], given by

$$s_{AX}(t_1,t_2) = C M_{oA}\cos(\Omega_A t_1)\exp(i\Omega_X t_2). \qquad [2.21]$$

As is clear from Eq. [2.21] the amplitude of the detected X-spin magnetization is modulated with frequency Ω_A as a function of t_1. The amplitude modulation is readily converted into phase modulation by stepping the phases of the r.f. pulses in Fig.2.29 as indicated in Table 2.2, analogous to the quadrature Jeener experiment described in section 2.3.2. The phase-modulated signal is then described by

$$s_{AX}(t_1,t_2) = -iC M_{oA}\exp(-i\Omega_A t_1)\exp(i\Omega_X t_2). \qquad [2.22]$$

A two-dimensional Fourier transformation will then result in a spectrum with cross peaks at (Ω_A,Ω_X) if the constant C is not equal to zero, i.e. if cross relaxation took place during the mixing period Δ. In order to eliminate correlation through scalar coupling as discussed in section 2.3, a field gradient pulse is applied during the mixing period Δ, destroying all coherences of order greater than zero.

It is in practice very hard to get quantitative results for relaxation or exchange rates from this experiment. A practical problem in performing the experiment is a proper choice for the length of the delay Δ of the mixing period. Looking at large peptides, a good approach for this problem appears to be to perform several experiments with a variety of values, e.g. 0.05, 0.1 and 0.3 sec. The first spectrum will then give only cross peaks for nuclei with a high cross relaxation rate (internuclear distance 2 - 3 Å) while the

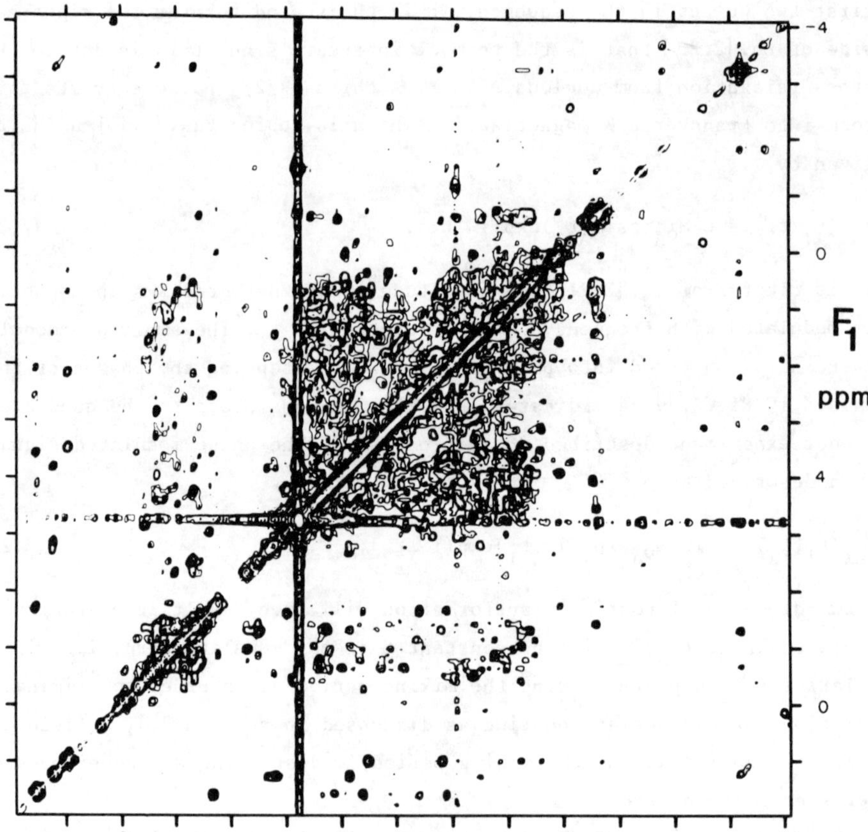

Fig.2.30 Shift correlation spectrum based on cross-relaxation of ferrocytochrome c. From ref.(43).

experiments with longer Δ delays will give a larger number of cross peaks due to the fact that lower cross relaxation rates, corresponding to internuclear distances for protons of up to 5 Å, also give cross peaks of significant intensity (41).

Since no opposite intensities occur within the cross multiplets, as present in the homonuclear scalar coupling correlation spectroscopy (section 2.3.1), no destructive interference in the cross multiplet takes place, and no need exists to resolve the individual multiplets in the 2D spectrum.

Looking at smaller, organic molecules, the choice for the length Δ of the mixing period will be in the order of the longitudinal relaxation time T_1 of the nuclei involved (42).

In principle the experiment can also be used in the heteronuclear case, investigating e.g. cross relaxation between ^{13}C and ^{1}H. No results obtained by the heteronuclear version of this experiment have been published until to date.

Fig.2.30 shows a typical example of a ^{1}H 2D spectrum obtained with the cross relaxation based correlation spectroscopy. The molecule under investigation is ferrocytochrome c, a biomolecule with an M.W. of about 12,500.

References

(1) A.A. Maudsley and R.R. Ernst, Chem. Phys. Lett. 50, 368 (1977).

(2) G. Bodenhausen and R. Freeman, J. Magn. Reson. 28, 471 (1977).

(3) A.A. Maudsley, L. Müller, R.R. Ernst, J. Magn. Reson. 28, 463 (1977).

(4) R. Freeman and G.A. Morris, J. Chem. Soc. Chem. Commun., 684 (1978).

(5) D.P. Burum and R.R. Ernst, J. Magn. Reson. 39, 163 (1980).

(6) M.H. Levitt and R. Freeman, J. Magn. Reson. 34, 675 (1979).

(7) A. Bax and G.A. Morris, J. Magn. Reson. 42, 501 (1981).

(8) D.I. Hoult and R.E. Richards, Proc.Roy. Soc. London A.344, 311 (1975).

(9) J.S. Waugh, J. Mol. Spectrosc. 35, 298 (1970).

(10) G. Bodenhausen and D.J. Ruben, Chem. Phys. Lett. 69, 185 (1980).

(11) A.F. Mehlkopf, Thesis, T.H. Delft 1978.

(12) G. Bodenhausen, J. Magn. Reson. 39, 175 (1980).

(13) G.A. Morris, J. Magn. Reson. 44, 277 (1981).

(14) J. Jeener, Ampere International Summer School, Basko Polje, Yugoslavia 1981.

(15) W.P. Aue, E. Bartholdi and R.R. Ernst, J. Chem. Phys. 64, 2229 (1976).

(16) W.A. Anderson and R. Freeman, J. Chem. Phys. 37, 85 (1962).

(17) A. Bax and R. Freeman, J. Magn. Reson. 44, 542 (1981).

(18) K. Nagayama, K. Wüthrich and R.R. Ernst, Biochem. Biophys. Res. Comm. 90, 305 (1979).

(19) A. Bax, R. Freeman and G.A. Morris, J. Magn. Reson. 42, 164 (1981).

(20) K. Nagayama, A. Kumar, K. Wüthrich and R.R. Ernst, J. Magn. Reson. 40, 321 (1980).

(21) W.A. Anderson, R. Freeman and C.A. Reilly, J. Chem. Phys. 39, 1518 (1963).

(22) G. Bodenhausen and P. Bolton, J. Magn. Reson. 39, 399 (1980).

(23) M. Tinkham, Group Theory and Quantum Mechanics, chapter 2, McGraw-Hill, New York 1964.

(24) A. Bax, P.G. de Jong, A.F. Mehlkopf and J. Smidt, Chem. Phys. Lett. 69, 567 (1980).

(25) I.D. Campbell, C.M. Dobson, R.J.P. Williams and A.V. Xavier, J. Magn. Reson. 11, 172 (1973).

(26) A. Bax, G.A. Morris and R. Freeman, J. Magn. Reson. 43, 333 (1981).

(27) G. Bodenhausen and R. Freeman, J. Magn. Reson. 36, 221 (1979).

(28) S. Vega and A. Pines, J. Chem. Phys. 66, 5624 (1977).

(29) S. Vega, J. Chem. Phys. 68, 5518 (1978).

(30) A. Wokaun and R.R. Ernst, J. Chem. Phys. 67, 1752 (1977).

(31) D.F. Evans and J.P. Maher, Proc. Chem. Soc. (London) 208 (1961).

(32) R. Freeman and D.M. Whiffen, Mol. Phys. 4, 321 (1961).

(33) R. Freeman, K.A. McLauchlan, J.I. Musher and K.G.R. Pachler, Mol. Phys. 5, 321 (1962).

(34) K.A. McLauchlan and D.H. Whiffen, Proc. Chem. Soc. (London), 144 (1962).

(35) A. Bax, A.F. Mehlkopf and J. Smidt, J. Magn. Reson. 35, 167 (1979).

(36) W.P. Aue, J. Karhan and R.R. Ernst, J. Chem. Phys. 64, 4226 (1976).

(37) G. Bodenhausen, R. Freeman, G.A. Morris and D.L. Turner, J. Magn. Reson. 31, 75 (1978).

(38) G. Bodenhausen, R. Freeman and D.L. Turner, J. Magn. Reson. 27, 511 (1977).

(39) J. Jeener, B.H. Meier, P. Bachmann and R.R. Ernst, J. Chem. Phys. 71, 4546 (1979).

(40) B.H. Meier and R.R. Ernst, J. A. Chem. Soc. 101,6441 (1979).

(41) A. Kumar, R.R. Ernst and K. Wüthrich, Biochem. Biophys. Res. Commun. 95, 1 (1980).

(42) S. Macura and R.R. Ernst, Mol. Phys. 41, 95 (1980).

(43) R. Baumann, A. Kumar, R.R. Ernst and K. Wüthrich, J. Magn. Reson. 44, 76 (1981).

3 J-SPECTROSCOPY

3.1 Introduction

In principle J-spectroscopy can be considered as a special case of chemical shift correlation spectroscopy. However, because the spectra obtained with J-spectroscopy generally show features quite different from the spectra described in the previous chapter, and because the theory is usually described differently too, this class of experiments is presented in a separate chapter.

In a conventional NMR spectrum of a molecule in a liquid, the positions of resonance lines are usually determined by two parameters, the chemical shift δ of the nucleus involved and the scalar interaction J with other nuclei in the same molecule. All two-dimensional J-spectroscopy experiments have in common that these two parameters, or linear combinations of them, are separated along the two axes of the two-dimensional spectrum, enabling a direct and unambiguous determination of both J and δ. Multiplets which are overlapping in the conventional spectrum can be separated in the 2D spectrum if the corresponding chemical shifts are different. Many experimental schemes have been proposed for both homonuclear and heteronuclear coupled spin systems (1-14). Most of these schemes create a spin echo by means of a π pulse. Therefore first a simple explanation of the spin echo effects, i.e. refocussing and echo modulation will be given.

Throughout the whole chapter it will be supposed that coupling between all nuclei is weak, and again relaxation is neglected, unless stated otherwise.

As was already shown by Hahn (15) in 1950, spin echoes occur if two or more radiofrequency pulses, spaced by a time delay, are applied to a spin system in an inhomogeneous static magnetic field. As an example the effect of a $(\pi/2)_x - t_1/2 - (\pi)_y - t_1/2$ sequence (Fig.3.1) applied to a set of isolated spins in thermal equilibrium will be explained.

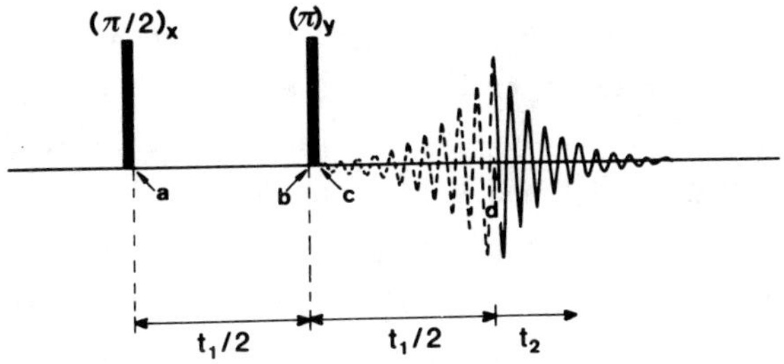

Fig. 3.1 Sequence generating a spin echo at a time t_1 after the initial $\pi/2$ pulse. This sequence is the basic scheme of homonuclear J-spectroscopy.

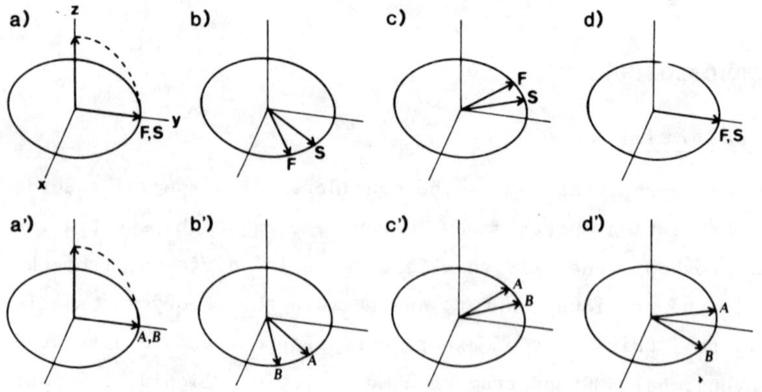

Fig. 3.2 The rotation during the sequence of Fig.3.1 of two isochromats of a sample consisting of non-coupled spins (a-d) in an inhomogeneous field and (a'-d') the rotation of the two components of a homonuclear doublet in a homogeneous magnetic field. The figures a-d correspond to the times a-d indicated in Fig.3.1. The isochromats are refocussed along the y-axis at time d, while the doublet components then have phases $\pm\pi J t_1$.

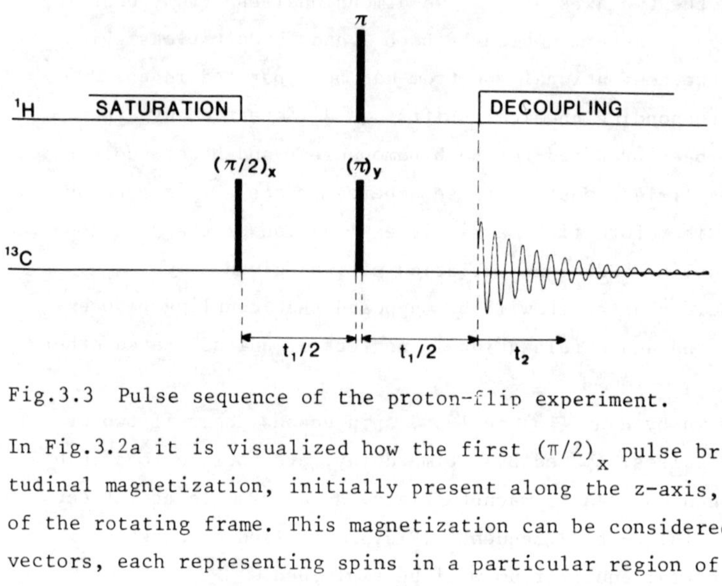

Fig.3.3 Pulse sequence of the proton-flip experiment.

In Fig.3.2a it is visualized how the first $(\pi/2)_x$ pulse brings the longitudinal magnetization, initially present along the z-axis, into the y-axis of the rotating frame. This magnetization can be considered as a sum of vectors, each representing spins in a particular region of the sample, with a characteristic strength of the magnetic field. In Hahn's nomenclature each of these vectors is an isochromat. During the period $t_1/2$ these isochromats diverge, since each has a slightly different precession frequency, due to the inhomogeneity of the static magnetic field. This is indicated in Fig.3.2b by a fast and a slow isochromat, labelled F and S. The $(\pi)_y$ pulse at time $t_1/2$ rotates each isochromat through an angle π about the y-axis. This means that now the fast isochromat lies behind the slow (Fig.3.2c). Assuming the position of both corresponding spins to be unchanged (i.e. no diffusion), and thus their difference in frequency still the same, it takes another time $t_1/2$ for the fast isochromat to catch up with the slow one. All isochromats will then be aligned along the y-axis again at time t_1 (Fig.3.2d), thus forming a spin echo. Thus, refocussing occurs and effects of chemical shifts and inhomogeneity of the static magnetic field are eliminated at time t_1 after the $(\pi/2)_x$ pulse. Due to transverse relaxation, the magnitude of the magnetization will have decreased a factor $\exp(-t_1/T_2)$ compared with the initial magnetization.

As pointed out by Hahn and Maxwell in 1952(16), in the case of a homonuclear coupled spin system, the echoes will be modulated by homonuclear coupling. This modulation effect will be briefly discussed below. As the effect of refocussing due to a π pulse has been discussed above, in the following discussion the effect of field inhomogeneity will be neglected. Consider for example a homonuclear system of two coupled nuclei with chemical shift frequencies $\Omega_A/2\pi$ and $\Omega_X/2\pi$ and mutual coupling J. The coupling is assumed to be weak, i.e. $(\Omega_A - \Omega_X) >> 2\pi J$. The effect of the sequence of Fig.3.1 on the two magnetization components of the doublet of nucleus A, labelled A and B, corresponding to respectively state α and β of the X-nucleus, will now be analyzed. Just before the $(\pi)_y$ pulse the two components will have positions as indicated in Fig.3.2b'. The $(\pi)_y$ pulse rotates the magnetization components to mirror image positions with respect to the y-axis. Apart from that, the $(\pi)_y$ pulse has an additional effect: the state of the coupled spin X is inverted. The component A rotating with frequency $(\Omega_A/2\pi - J/2)$ before the $(\pi)_y$ pulse, will be coupled to a spin in state β after the $(\pi)_y$ pulse and then precess with frequency $(\Omega_A/2\pi + J/2)$, i.e. the frequency of component B (Fig.3.2c'). At time t_1 (Fig.3.2d') this doublet component will then have a phase $\pi J t_1$, and similarly the other component will have a phase $-\pi J t_1$. The multiplet components are thus modulated in phase as a function of t_1. This is the basis of most experiments discussed in this chapter.

The echo modulation only occurs in the case of homonuclear coupling; in the case of heteronuclear coupling, the second nucleus coupled to the one under observation, will not flip its state. In the heteronuclear case the echo modulation can be restored by applying a π pulse to this nucleus too, as shown in Fig.3.3.

3.2 Heteronuclear J-spectroscopy

3.2.1 The proton-flip experiment

The proton-flip experiment has been proposed by Bodenhausen et al. (4) and Kumar et al. (5) and its experimental aspects have been described in a series of papers (6-8, 17). The method employs the pulse sequence of Fig.3.3. In principle the experiment can be applied to any set of heteronuclear coupled spins, but until now it has been mainly used for studying spin systems consisting of coupled protons and carbon-13 nuclei, observing the ^{13}C signal. One might expect the spin echoes in this case to be modulated by both homo- and heteronuclear coupling, but since the ^{13}C nucleus is quite rare, the homonuclear modulation is almost absent in this case, and will be neglected in the following discussions.

The frequency of an arbitrary ^{13}C multiplet component can always be written as $\Omega/2\pi + \sum_i m_i J_i$, where $\Omega/2\pi$ denotes the chemical shift frequency, m_i the magnetic quantum number of proton i and J_i the coupling constant with proton i. Just before the ^{13}C $(\pi)_y$ pulse in Fig.3.3, magnetization corresponding to this multiplet component will have accumulated a phase ϕ given by

$$\phi = (\Omega + 2\pi \sum_i m'_i J_i) t_1/2 \qquad [3.1]$$

where m'_i denotes the magnetic quantum number of proton i before the proton π pulse.

The angle ϕ is inverted by the ^{13}C $(\pi)_y$ pulse and the magnetic quantum number of the protons is inverted by the proton π pulse. The new frequency of the magnetization considered after the proton π pulse is then equal to $\Omega/2\pi + \sum_i m_i J_i$, with $m_i = -m'_i$.

At time t_1 in the pulse sequence of Fig.3.3 this component will have accumulated a phase $2\pi t_1 \sum_i m_i J_i$. If broad-band decoupling is started at time t_1, this component will continue rotating with angular frequency Ω and the time domain signal for this component is given by

$$s(t_1,t_2) = M_o \exp(2\pi i \sum_i m_i J_i t_1) \exp(i\Omega t_2) \qquad [3.2a]$$

Clearly this is a case of phase modulation (section 1.5.1), from which a two-dimensional Fourier transformation will generate a phase-twisted resonance at position $(\Omega/2\pi, \sum_i m_i J_i)$. As shown below the signal can be considered as being modulated in amplitude, enabling the recording of 2D spectra in the double absorption mode.

In the case where the coupling among the (non-equivalent) protons is weak, the ^{13}C multiplet structure will be symmetric and there will always be a counter-partner for each multiplet component of equal magnitude M_o coupled to protons i' with magnetic quantum numbers opposite to those of the original component $(m'_i = -m_i)$. The time domain signal for this component is described by

$$s'(t_1,t_2) = M_o \exp(-2\pi i \sum_i m_i J_i t_1) \exp(i\Omega t_2) \qquad [3.2b]$$

The sum of both components is given by

$$s^+(t_1,t_2) = 2M_o \cos(2\pi \sum_i m_i J_i t_1) \exp(i\Omega t_2) \qquad [3.3]$$

which represents a signal modulated in amplitude.

As explained in section 1.5.2 a two-dimensional Fourier transformation can then generate a spectrum in the double absorption mode. The position of a resonance along the F_2-axis is determined by the ^{13}C chemical shift, while along the F_1-axis the multiplet structure appears. Apart from digitization problems the resonance lines will have natural width along the F_1-axis because, neglecting diffusion, static magnetic field inhomogeneity is eliminated by the refocussing pulse applied to the ^{13}C at time $t_1/2$. Line widths in the F_2-dimension are mainly determined by transverse relaxation, inhomogeneity broadening and poor decoupling.

For large spin systems the transverse relaxation times of coupled spins are often short because of scalar relaxation through the protons (18), making it impossible to resolve the indirect ^{13}C-^{1}H couplings. Line-widths are apparently increased by very small unresolved long-range couplings. The poor resolution due to scalar relaxation and unresolvable long-range couplings will be identical for the multiplet structure in the two-dimensional spectrum.

In the case where the coupling between the protons is strong, the simplified discussion about echo-modulation given in section 3.1 is not valid (17). In this case the stationary wave functions of the spin system are not simply given by basic product functions of the individual spins, but by linear combinations of these (section 3.3.3). A spin inversion of all the protons by means of a π pulse causes a stationary wave function, describing a certain state, to become a linear combination of stationary wave functions. This implies that a certain multiplet component before the proton π pulse will be split into several components rotating with different frequencies (section 3.3.3). A much more complicated fine structure of the multiplet will occur along the F_1-axis of the 2D spectrum in this case. A detailed analysis of this effect is given in ref.(17).

3.2.2 The gated-decoupler experiment

Apart from the proton-flip, another method has been proposed by Bodenhausen et al. (3), which is based on a similar echo-modulation. The pulse scheme of this so-called gated-decoupler experiment is shown in Fig.3.4. This method differs from the proton-flip experiment in the fact that during the first half of the evolution period broad-band decoupling of the protons is applied and no ^{1}H pulse. In Fig.3.5 the behaviour of the two components of a ^{13}C doublet during the sequence is shown, while for simplicity field inhomogeneity is neglected.

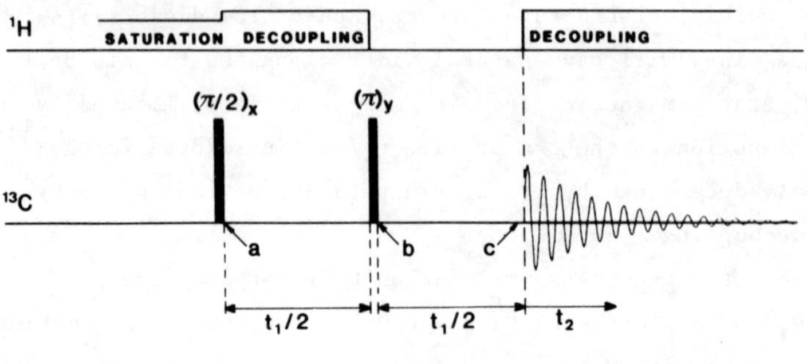

Fig.3.4 Pulse scheme of the gated-decoupler experiment.

The $(\pi/2)_x$ pulse turns both multiplet components along the y-axis (Fig.3.5a), then during the first half of t_1 they will both precess with the same chemical shift $\Omega/2\pi$ since proton decoupling is applied during that period. At time $t_1/2$ the $(\pi)_y$ pulse to the ^{13}C inverts the phase $\Omega t_1/2$ of both components (Fig.3.5b). Then for a time $t_1/2$ no decoupling is applied and the components rotate with angular frequencies $\Omega \pm \pi J$ and have accumulated phases $\pm\pi J t_1/2$ at time t_1 (Fig.3.5c). Since from then on proton decoupling is switched on again, both components will precess during the detection period t_2 with identical angular frequencies Ω. Just like in the previous section (Eqs.[3.1]-[3.3]) the detected signal can be considered as either originating from a single component modulated in amplitude by $\cos(\pi J t_1/2)$ or from the sum of two components with opposite phase modulation. As will be shown at the end of this section, this last picture will be helpful in analysing the effect of strong coupling among protons, possibly present in the case of a more complicated spin system.

In the case of weak coupling the assumption of pure amplitude modulation is justified and a two-dimensional absorption spectrum can be obtained in the usual fashion (section 1.5.2). Because the frequency of the amplitude modulation in this experiment is equal to only half of the modulation frequency in the proton-flip experiment, the multiplet lines will be separated along the F_1-axis only by half the coupling constant, therefore in principle

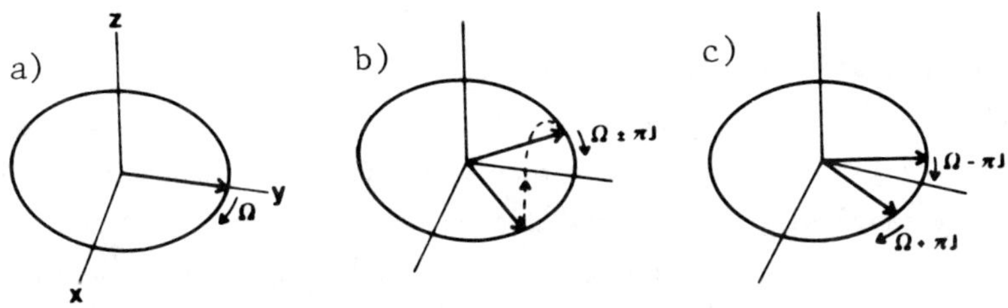

Fig.3.5 Rotation of the two components of a carbon-13 doublet during the gated-decoupler sequence of Fig.3.4 (see text).

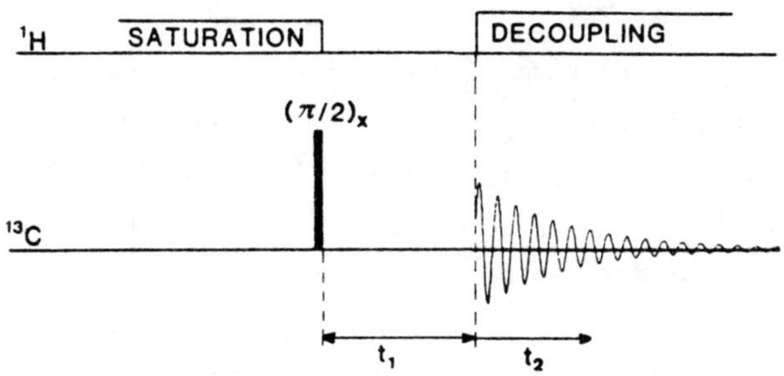

Fig.3.6 Experimental scheme of the foldover gated-decoupler experiment.

decreasing the attainable resolution by a factor two (see section 3.2.4).

The effect of strong coupling among the protons

The effect of strong coupling among protons in a molecule where a ^{13}C nucleus is present* on the eventual two-dimensional spectrum can be analyzed in a simple way. In this case the ^{13}C multiplet structure can be asymmetrical about the chemical shift frequency. As a consequence, at the end of the evolution period there are no pairs of magnetization components with exactly opposite phases with respect to the y-axis. Therefore in this case the experiment cannot be treated as amplitude modulation, but each multiplet component has to be considered separately as a phase-modulated signal. The frequency of this modulation equals half the difference in frequency between the multiplet component considered and its chemical shift frequency. This implies that the multiplet structure along the F_1-axis in the case of strong coupling is identical to the multiplet structure in a conventional experiment. Because of the phase modulation no absorption spectra can be obtained in this case (section 1.5.1). However, a single cross-section through the 2D spectrum parallel to the F_1-axis at the chemical shift frequency of a certain carbon-13 can be phased just like conventional one-dimensional spectra to yield a pure absorption multiplet structure. Freeman et al. (12) developed a modification on the gated-decoupler experiment which does make it possible to produce a complete two-dimensional absorption spectrum in the case of strong coupling.

3.2.3 The foldover gated-decoupler experiment

One of the earliest two-dimensional experiments proposed in literature (1975) was a gated-decoupler experiment without a refocussing pulse (1) (Fig. 3.6). It was much later before the practical use of this experiment was pointed out by Müller (14). The theory of this experiment is quite analogous to the normal gated-decoupler experiment. During the time t_1 the proton-coupled magnetization vectors rotate with their characteristic frequency, determined by both chemical shift and J-coupling. Then during the detection period broad-band decoupling of the protons is applied, and all ^{13}C multiplet components will precess during that period with the chemical shift frequency. The vector sum of all multiplet components will thus be modulated as a function of t_1, both in amplitude and in phase (section 1.5.3). However,

---

* These are the protons giving rise to the carbon-13 satellites in conventional proton spectra.

the detected signal can also be considered as the sum of a set of components, each component being only modulated in phase with its precession frequency during the evolution period. Therefore phase modulation of the detected signal occurs with the frequencies of the proton coupled spectrum. Hence, the chemical shift frequency is now present along both axis of the 2D spectrum, and a high sampling frequency along the t_1-axis is needed to avoid folding, inferring a very large data matrix. Müller proposed a solution for this problem by allowing a lower sampling frequency along the t_1-axis and correcting then for the consequent folding errors by a data manipulation trick (14). Another possibility to avoid the folding problem altogether and still allowing a low sampling frequency is presented in ref. (19).

3.2.4 Comparison of methods

In Table 3.1 some properties of the three different experiments are shown. Which experiment should be chosen in a particular case depends mainly on the multiplet resolution required. Two different cases can be distinguished: First the study of the fine structure of the multiplet, for the investigation of long-range couplings and second, the determination of only the multiplicity, i.e. the number of protons directly bonded to a carbon-13.

Table 3.1. Comparison of the properties of the proton-flip, the gated-decoupler and the foldover gated-decoupler experiments.

	proton-flip	gated-decoupler	foldover gated-decoupler
Attainable resolution	good	medium [1]	medium [1]
2D Absorption	yes	yes [2]	no
Critical settings [3]	yes	medium	no
Simplicity of spectrum	medium [4]	good	good

[1] Depending on the spectrometer (see text)

[2] Depending on the problem (see section 3.2.2)

[3] To avoid artefacts in the 2D spectrum (see section 3.4)

[4] Only in the case of strong coupling among the protons (see text)

Fine structure

As can be seen in Table 3.1, in the case of weak coupling among the protons the proton-flip experiment should be preferred, because of the highest resolution along the F_1-axis. In the case of strong coupling among the protons the different and sometimes complicated multiplet structure obtained with the proton-flip experiment can be confusing, and it can be better to use the gated-decoupler or the foldover gated-decoupler experiment. Comparing these last two methods does not give exclusive preference to one of them. In both experiments it is possible to phase the appropriate multiplet cross-section to the absorption mode. The line-width along the F_1-axis in the foldover gated-decoupler experiment is equal to $(\pi T_2^{\dagger})^{-1} + (\pi T_{2_{cpl}})^{-1}$, where $T_{2_{cpl}}$ denotes the transverse relaxation time of the coupled resonances, and an extra field inhomogeneity decay with time constant $T_2^{\dagger}$ is assumed. In the gated-decoupler experiment the width is given by $(\pi T_{2_{dec}})^{-1} + (\pi T_{2_{cpl}})^{-1}$, where $T_{2_{dec}}$ describes the relaxation of the decoupled carbon-13 resonance. Hence, if $T_{2_{dec}} > T_2^{\dagger}$ the normal gated-decoupler experiment should be preferred.

Because of the poor sensitivity in the application to high multiplet resolution, acquisition time along the t_2-axis should at least be equal to the decay constant $T_2^{\star}$ of the decoupled resonances. Because a high resolution along the F_1-axis is wanted, a long acquisition time along the t_1-axis is required too. In practice these two facts will often lead to a very large data matrix.

Multiplicity

If the only purpose is to determine the number of protons directly bonded to a ^{13}C only a low resolution of the multiplet structure (20-40 Hz) is required. A sampling time along the t_1-axis of 25 to 50 msec. for the proton-flip and foldover gated-decoupler experiment, and 50 to 100 msec. for the normal gated-decoupler experiment is usually sufficient. Choosing a longer sampling time along the t_1-axis will unnecessarily lengthen the experiment, use a larger data matrix and decrease the sensitivity.

Consulting Table 3.1 shows that in this application the gated-decoupler experiment is the best one to use.

In favourable cases, the sensitivity of the experiment is higher than for a conventional proton-coupled one-dimensional ^{13}C spectrum, since in the two-dimensional experiment the decoupled resonances are recorded, which often decay much slower than the proton-coupled ^{13}C signal.

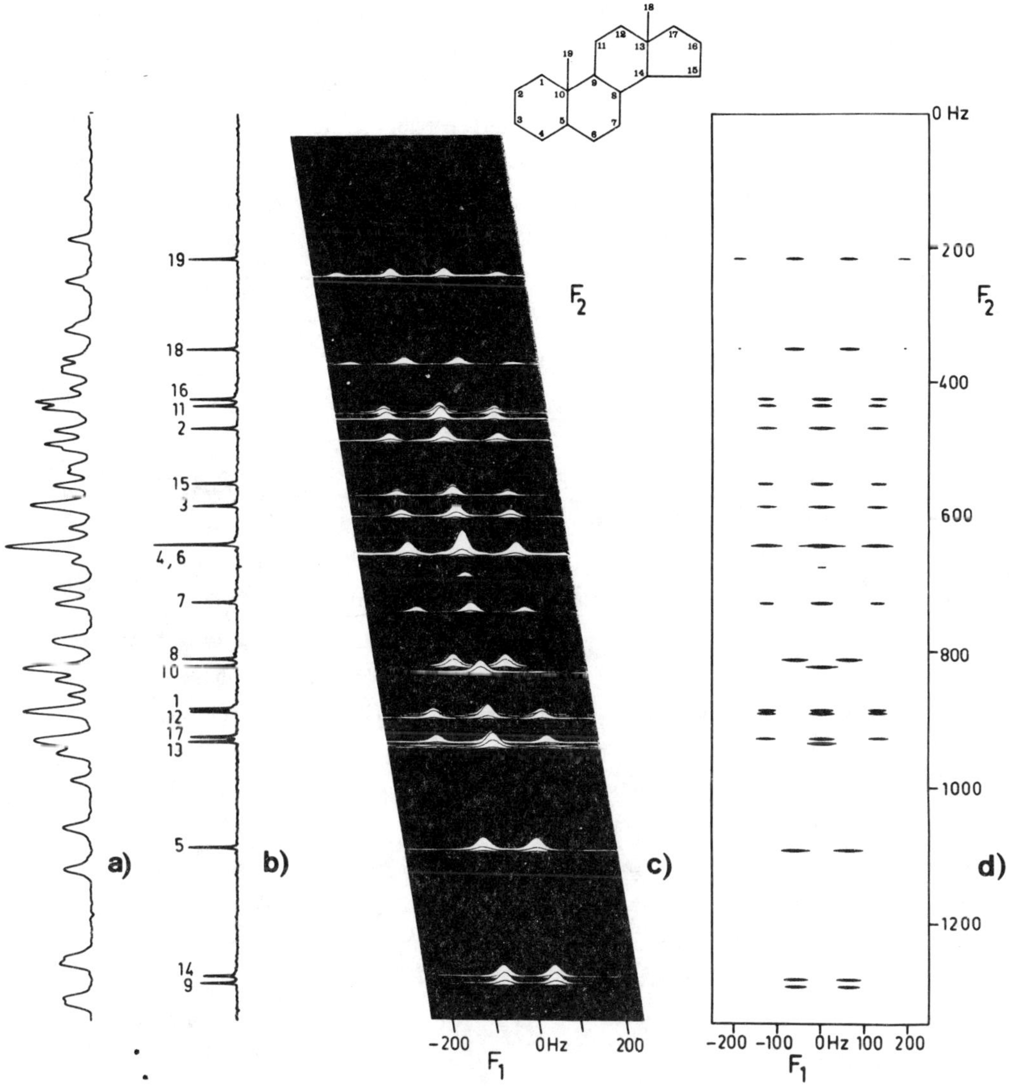

Fig.3.7 The conventional one-dimensional carbon-13 spectra of 5α-androstane at 50 MHz (a) proton-coupled and (b) proton-decoupled.
(c) A stacked trace plot and (d) a contour plot of the two-dimensional J-spectrum obtained with the gated-decoupler sequence.

Because the ^{13}C multiplet width is usually smaller than 500 Hz, a sampling frequency along the t_1-axis of 500 Hz (Δt_1= 2msec.) is usually sufficient for the proton-flip and foldover gated-decoupler experiment, while only 250 Hz (Δt_1= 4msec.) is needed in the normal gated-decoupler experiment. This means that usually 12 to 25 different t_1-values are sufficient to obtain the required information. As a consequence only relatively small data matrices are needed in this case.
As an illustration the normal gated-decoupler experiment is applied to 5α- androstane. The data were obtained from an experiment with 24 t_1-increments, lasting 24 x 3 seconds. In Fig.3.7 both a contour plot and a stacked trace plot are shown. Making the contour plot takes less time but can easily miss the outer lines of the quartets if the treshold is not properly chosen. As a comparison Fig.3.7 also shows the conventional ^1H-coupled spectrum, obtained in the same measuring time.

3.3 Homonuclear J-spectroscopy

In homonuclear J-spectroscopy there exists only one basic scheme (2) on which a few modifications have been proposed (9-11, 13). The basic pulse scheme is shown in Fig.3.1. As has been explained in the introduction of this chapter, the phase of the magnetization components of a doublet at time t_1 equals $\pm \pi J t_1$. More generally the phase of a multiplet component at frequency $\Omega/2\pi + \sum_i m_i J_i$ is given by $2\pi \sum_i m_i J_i t_1$. This means phase modulation at a frequency $\sum_i m_i J_i$, and therefore a two-dimensional Fourier transformation of the time domain signal from this component will give a resonance line in the two-dimensional spectrum at $(2\pi \sum_i m_i J_i, \Omega + 2\pi \sum_i m_i J_i)$ if $\Omega/2\pi$ is the chemical shift frequency of the nucleus involved. The peak positions for a proton AX_2 spin system in a two-dimensional J-spectrum are schematically shown in Fig.3.8a. All resonance lines of multiplet with chemical shift frequency $\Omega/2\pi$ lie on a cross-section which makes an angle of 45° with the F_2-axis. A projection of such a spectrum in the absolute value mode★ at 45° on the line $F_1 = 0$ will yield a spectrum which only con-

---

★ The cross-section projection theorem (21) states that the projection of a phase-sensitive 2D spectrum on a line which makes an angle α with the positive F_2-axis (Fig.3.9a) is equal to the Fourier transform of the cross-section through the corresponding time-domain data at the same angle α (Fig.3.9b). Because only data are acquired for $t_1 > 0$ and $t_2 > 0$, the cross-section through the time domain data will contain only zeroes. Hence, a projection at 45° of a phase-sensitive 2D J-spectrum will not give a decoupled spectrum. Nagayama et al. (22) were the first to explain these effects.

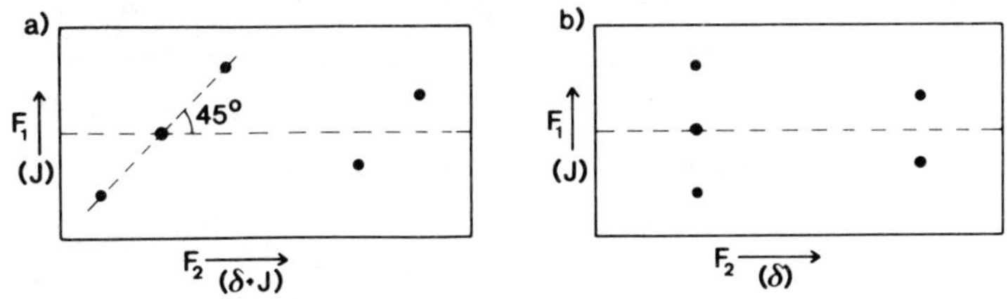

Fig.3.8 Schematic two-dimensional homonuclear J-spectrum of an AX_2 spin system (a) normal and (b) tilted. The angle of 45° in the non-tilted spectrum looks to be close to 90° in practice because the frequency scales along the two axes are usually different.

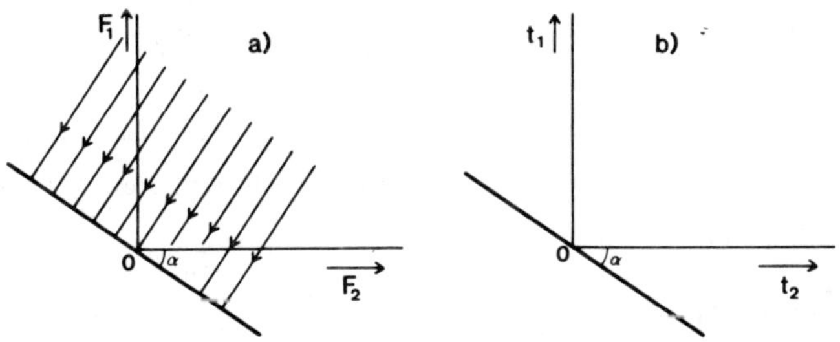

Fig.3.9 Illustration of the cross section projection theorem. The projection of the phase-sensitive 2D spectrum (a) on a line which makes an angle α with the positive F_2-axis is equal to the Fourier transform of a cross section (b) through the time domain which makes the same angle α with the t_2 axis.

tains peaks at chemical shift frequencies, a so-called broad-band homonuclear decoupled spectrum. As explained in the footnote, the projection of a phase-sensitive spectrum at this angle always gives zero.

A tilting of the spectrum according to $(\omega_1', \omega_2') = (\omega_1, \omega_2 - \omega_1)$ gives, as schematically visualized in Fig.3.8b, a spectrum in which both J and Ω are completely separated. The tilting is usually done by means of a computer routine which calculates the new data matrix via shearing of the data in the original matrix. A cross-section through the tilted spectrum parallel to the F_1-axis for a chemical shift frequency of F_2, will show the multiplet structure with a high resolution. This high resolution is analogous to heteronuclear J-spectroscopy mainly due to the fact that the π pulse eliminates the effects of static magnetic field inhomogeneity during t_1.

Because the method is based on phase-modulation the resonance lines have the phase-twisted shape (section 1.5.1). A resonance line on a cross-section

parallel to one of the main axes in the non-tilted spectrum can always be phase-corrected to the pure absorption mode. Unfortunately this is impossible in the case of the tilted spectrum if a cross-section parallel to the F_1-axis is taken. This can be understood by considering the expression for a phase-twisted resonance line as given in section 1.5.1 (Eq.1.48). However, an absorption-like spectrum can be obtained, with significant distortions at the base of the resonance line, wrongly identified by Sukumar and Hall (23) as t_1-noise. However, this phase-sensitive presentation still shows much better resolution of the multiplet structure than the absolute value mode (23).
As an example, in Fig.3.10 an absolute value mode 2D J-spectrum of the tricyclodecane derivative, sketched on the inset, is shown. Cross-sections taken through the phase-sensitive spectrum show the structure of some of the multiplets.

The actual technical approach of a problem, using two-dimensional J-spectroscopy depends largely on the nature of the problem. Three different cases will be discussed in the following sections.

3.3.1 Simplification of spectra

As shown above, a projection of the tilted 2D J-spectrum gives a homonuclear decoupled spectrum, while at the same time multiplet structures are available from the cross-sections of the spectrum parallel to the F_1-axis. The resulting simplification can in principle be quite large. However, as pointed out in ref. (22) the projected spectrum will have a low resolution. The half-width of a 45° projection of a Lorentzian line in the absolute value mode, symmetric in F_1 and F_2 appears to be a factor 4.5 broader than the corresponding line in a one-dimensional absorption spectrum (22). The lines in a projection show very strong tailing too. Therefore a real simplification can only occur if the conventional one-dimensional spectrum consists of overlapping but individually (partly) resolved multiplets. Examples of practical applications have been shown by Nagayama et al. (22, 24) and Morris, Hall and Sukumar (25, 26).

Because line shapes and line widths are the main limiting factors in the use of 2D J-spectroscopy, proper digital filtering is in practice very important. Some considerations and suggestions with respect to digital filtering will be given below. First a strategy will be described to get highest resolution in the projected decoupled spectrum. Because the projection is calculated in the absolute value mode, the absolute value line shape in the two-dimensional spectrum determines the eventual resolution in the decoupled

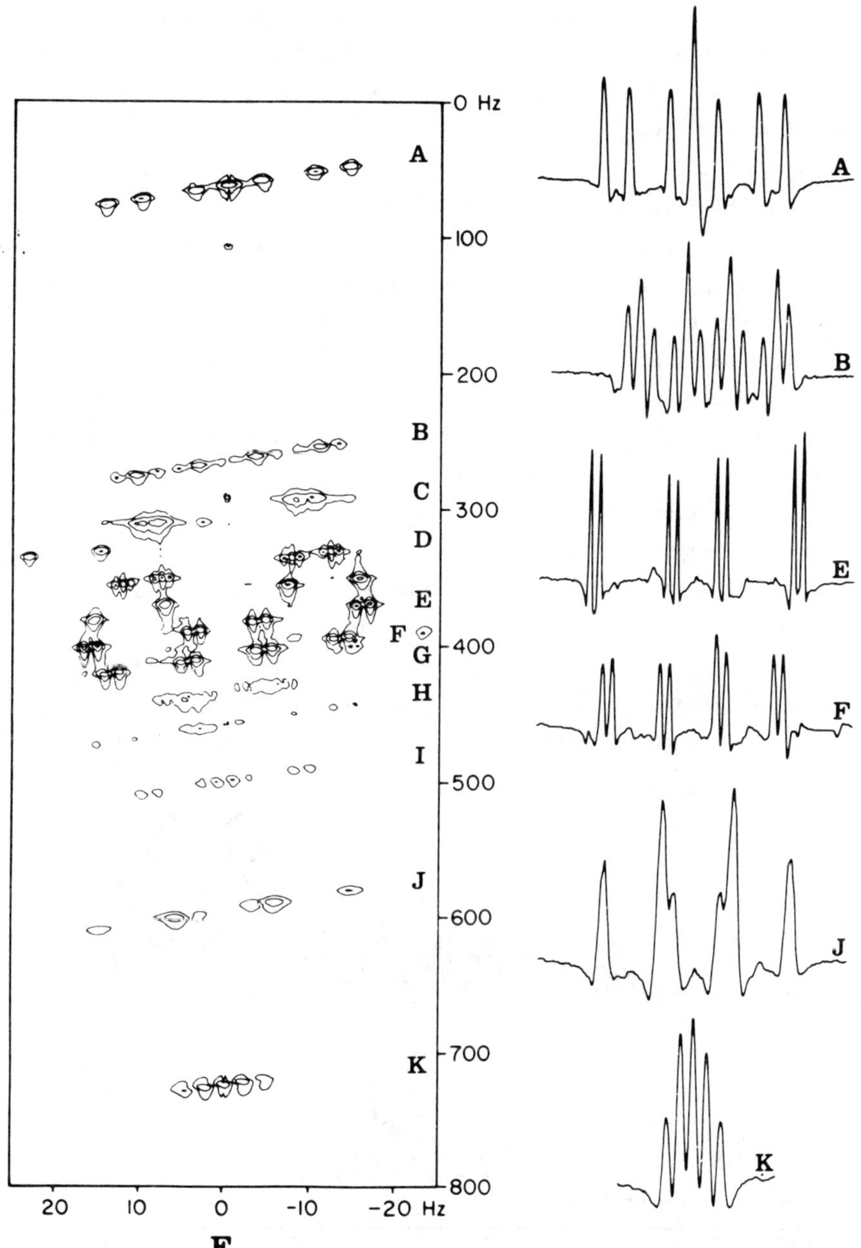

Fig.3.10 Two-dimensional proton J-spectrum of the tricyclodecane derivative shown in the inset in Fig.3.11, presented in the absolute value mode. At the right hand side some multiplets, obtained from cross sections through the phase-sensitive 2D J-spectrum are shown.(Spectrum recorded at 200 MHz).

spectrum. As explained in section 1.7.3 the pseudo-Gaussian echo filter will, if applied to both axes, produce line shapes which have absorptive character in the absolute value mode. Projections of such a 2D spectrum will then give a decoupled spectrum with absorptive Gaussian line shapes. The main practical disadvantages of this filter are the large intensity distortions and the sensitivity loss which it introduces. Therefore it is suggested to use the pseudo-echo filter only along the t_2-axis to remove the dispersion contribution from the line shape in the F_2-dimension, and to use the Lorentzian-Gaussian transform filter (section 1.7.2) for line-narrowing along the t_1-axis. The absolute value tailing in the F_1-dimension of the relatively narrow resonance lines in this dimension will in practice not severely distort the line shape of the projection. In this case real line-narrowing only takes place along the F_1-axis, while along the F_2-axis the line shape is modified to give less absolute value tailing. The possibilities of digital filtering of the two-dimensional time signal should not be overestimated: an analysis shows that in practice it is very hard to get a decoupled projection with lines as narrow as or narrower than the individual multiplet components in a conventional spectrum which has not been resolution enhanced. The limitations of the possible line-narrowing in the 2D spectrum is set by intensity distortion and sensitivity problems. Fig.3.11 shows the comparison of a conven-

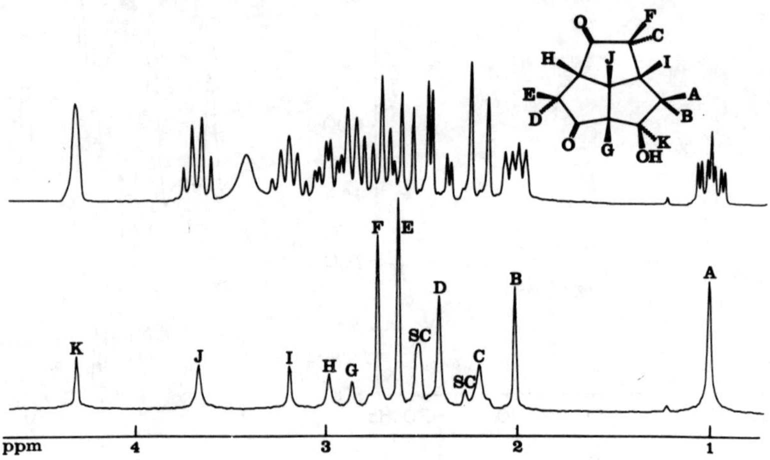

Fig.3.11 The conventional one-dimensional 200 MHz proton spectrum (top trace) of the tricyclodecane derivative shown in the inset. The position of the broad resonance of the OH depends on temperature and concentration. The bottom trace shows the 45° projection of the two-dimensional J-spectrum with pseudo Gaussian echo filtering used in the t_2 dimension to suppress dispersion mode contributions. Lines marked "SC" arise due to the strong coupling. From ref.(44)

tional ^1H spectrum and its decoupled equivalent obtained by using digital filtering as mentioned above. The reader is warned to be cautious in interpreting 2D J-spectra and 45° projections if there is strong homonuclear coupling. Extra resonances will appear, as demonstrated in section 3.3.3. Multiplet cross-sections can best be studied in the phase-sensitive mode and in this case it is better not to use the pseudo-echo filter, since this introduces large line shape distortions (see e.g. Fig.1.10c).

3.3.2 Improvement of multiplet resolution

In the case where the multiplet resolution is limited by static magnetic field inhomogeneity, 2D J-spectroscopy can be used to eliminate this inhomogeneity broadening. Several different one-dimensional methods having similar effects are also available (27, 28, 29). The actual experimental approach, using 2D J-spectroscopy, is slightly different from the one given in the previous section. Because the decay of the signal as a function of t_1 depends in first approximation on transverse relaxation only, natural line widths along the F_1-axis of the 2D spectrum can be obtained. In order to use additional line-narrowing along this axis, an acquisition time along the t_1-axis of up to eight times T_2 sometimes has to be chosen. Even though the spectral width along the F_1-axis is only determined by maximum multiplet width, this often results in such a large number of t_1-increments that the sampling time along the t_2-axis has to be limited because of the maximum size of the data matrix available.
The Lorentzian to Gaussian transformation (section 1.7.2) is recommended for line-narrowing along the t_1-axis, while along the t_2-axis Lorentzian or Gaussian weighting can be used to improve sensitivity and to avoid possible truncation effects. It is mentioned here that line-narrowing along the F_1-axis does not change the ratio of t_1-noise amplitude to signal amplitude.

In practice the effect of diffusion in the sample will prevent perfect refocussing, and a decay as a function of t_1 will occur faster than expected. As pointed out by Lösche (30) and suggested by Aue et al. (2) a series of equally spaced π pulses (Fig.3.12) during the evolution period can be used to eliminate the diffusion effect. As will be pointed out in section 3.5.1, new problems arise in this case due to cumulative effects of errors in the π pulse.

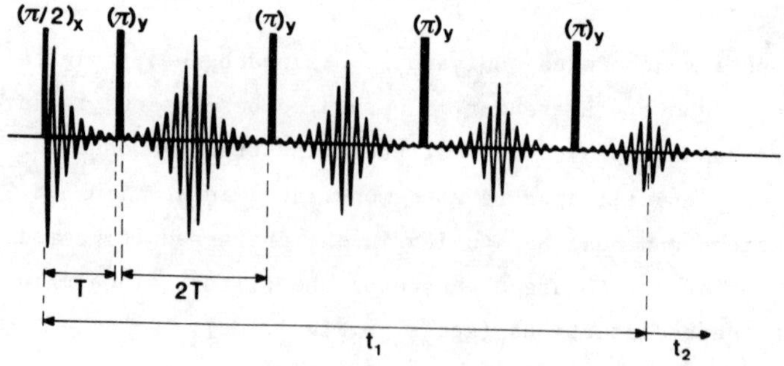

Fig.3.12 The elimination of diffusion effects by the application of a series of equally spaced π pulses in the evolution period.

3.3.3 Investigation of strong coupling

The semi-classical picture of homonuclear J-spectroscopy is not valid any more in the case where the nuclei are strongly coupled, i.e. where the mutual J-coupling is of the same order as the difference in chemical shift frequency, or even larger. Detailed investigations of the behaviour of several types of spin systems during the standard pulse sequence of Fig.3.1 have been made by Kumar (31) and Bodenhausen et al. (32). In this section a qualitative analysis is given of where the picture breaks down, and what the consequences for the 2D J-spectrum are. As an example the most simple case of strong coupling, an isolated AB spin system is chosen.

The reason for the extra complexity occuring in the case of strong coupling is that the eigenfunctions of the system are not given any more by simple basic product functions, but by linear combinations of these. The four stationary wave functions for the AB system are given in the second column of Table 3.2.

Table 3.2. The wave functions of an AB spin system before and after the spin inversion π pulse, expressed in the stationary wave functions. $\theta = 0.5 \arctan\{J/(\delta_A - \delta_B)\}$

Before the π pulse		After the π pulse	
short notation	stationary wave function	new state expressed in basic product functions	new state expressed in stationary wave functions
$\|1\rangle$	$\|\beta\beta\rangle$	$\|\alpha\alpha\rangle$	$\|4\rangle$
$\|2\rangle$	$\cos(\theta)\|\alpha\beta\rangle + \sin(\theta)\|\beta\alpha\rangle$	$\cos(\theta)\|\beta\alpha\rangle + \sin(\theta)\|\alpha\beta\rangle$	$\sin(2\theta)\|2\rangle + \cos(2\theta)\|3\rangle$
$\|3\rangle$	$\cos(\theta)\|\beta\alpha\rangle - \sin(\theta)\|\alpha\beta\rangle$	$\cos(\theta)\|\alpha\beta\rangle - \sin(\theta)\|\beta\alpha\rangle$	$\cos(2\theta)\|2\rangle - \sin(2\theta)\|3\rangle$
$\|4\rangle$	$\|\alpha\alpha\rangle$	$\|\beta\beta\rangle$	$\|1\rangle$

As a short notation for these eigenfunctions the state number as indicated in the first column is used. The effect of a π pulse is to transform the wave function of a spin in a certain eigenstate given in column 2, into a state given in the third column, which is the sum of stationary wave functions as indicated in the last column.

The effect of the standard pulse sequence of Fig.3.1 on one of the transitions of the AB spin system is shown in Fig.3.13. The $(\pi)_y$ pulse at time $t_1/2$ inverts the phase of the component as usual. As shown in Table 3.2 state $|1>$ converts completely into state $|4>$. However, state $|2>$ splits into a linear combination of $|2>$ and $|3>$. Hence, component ω_{12} will split into two components, rotating with angular frequencies ω_{24} and ω_{34}, and respective amplitudes $\sin^2(2\theta)$ and $\cos^2(2\theta)$ where θ is a measure for the strength of the coupling defined as $\theta = 0.5 \arctan\{J/(\delta_A - \delta_B)\}$. The new component rotating with ω_{34} can be considered as the normal component, while the ω_{24}-component is additional, due to the strong coupling. The phase of this component at time t_1 equals $(\omega_{24} - \omega_{12})t_1/2$ and it will cause a resonance line at $\{(\omega_{24} - \omega_{12})/2, \omega_{24}\}$ in the eventual 2D spectrum. For each of the three other components a similar splitting occurs. Hence, four additional lines will show up in the 2D spectrum.

An example of a 2D J-spectrum of a strongly coupled AB system ($\theta \simeq 0.2$ radian) is shown in Fig.3.14. The extra lines are on a cross-section which makes an

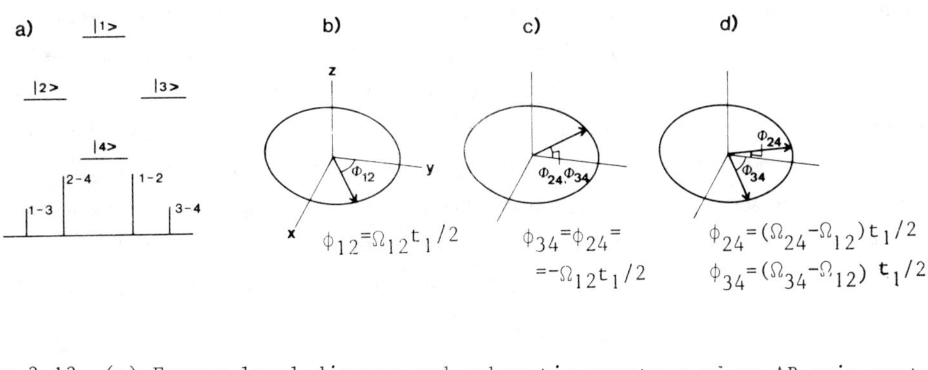

Fig.3.13 (a) Energy level diagram and schematic spectrum of an AB spin system. (b) Position of component 1-2 just before the $(\pi)_y$ pulse (Fig.3.1). (c) Positions of the two new components after the $(\pi)_y$ pulse and (d) positions at the end of the evolution period.

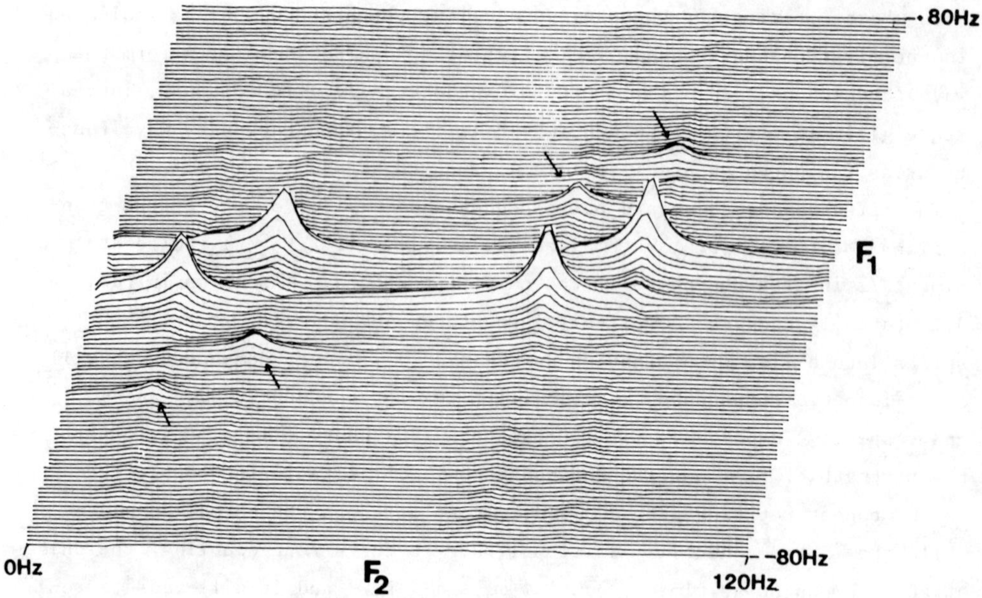

Fig.3.14 Example of an intermediately strongly coupled AB spin system ($\delta/J \approx 3.5$) recorded at 300 MHz. The four extra lines due to the strong coupling are indicated by arrows.

angle of 45° with the F_2-axis, cutting the line $F_1 = 0$ at the average frequency of the two chemical shifts involved. In the 45° decoupled projection these lines will give an extra resonance line just in the mid-point between the two real decoupled resonances. Note that these two real resonances do not occur at the exact chemical shift frequencies, but at the centers of the multiplets.

As is clear, the 2D J-spectrum for strongly coupled spin systems will be much more complicated than their weakly coupled equivalents. The number of resonance lines will be much larger than in the corresponding one-dimensional spectrum. Because of the spread of these resonance lines into two frequency dimensions, the spectrum will not necessarily be more crowded than the conventional spectrum and information can be extracted which would be hidden in the conventional spectrum. Because of the complicated nature of the two-dimensional spectrum in this case, information can generally only be extracted with the aid of a proper two-dimensional simulation program(32).

3.4 Artefacts in two-dimensional J-spectra

As mentioned in the introduction of this chapter, J-spectroscopy can be considered as a special case of correlation spectroscopy. The start of the acquisition is in these experiments delayed until the centre of the echo. The actual mixing pulse is now for most experiments the π pulse. Because of the peculiar properties of a π pulse, magnetization will only be transferred from one multiplet component to its mirror image with respect to the centre of the multiplet. However, as soon as the pulse is no longer exactly π radians, all other kinds of transfer will take place too, analogous to the homo- and heteronuclear shift correlation experiments described in the previous chapter. These other unwanted transfers give rise to the same kind of cross-peaks as in the 2D shift correlation spectra, and will usually be folded one or more times along the F_1-axis, because the F_1-spectral width in J-spectroscopy is lower than in the correlation experiments. These extra, unwanted lines, generally referred to as artefacts, can partly be eliminated by cycling the phases of pulses and receiver in a way as described by Bodenhausen et al. (33), which they called Exorcycle. Certain artefacts however, due to the non-exact inversion of the coupled spin cannot be eliminated this way (20). A good solution to suppress both these artefacts and the artefacts which can be suppressed with the Exorcycle is the use of a composite π pulse (34-36). This composite pulse consists of a $(\pi/2)_x-(\pi)_y-(\pi/2)_x$ pulse sequence and gives much better inversion than a normal π pulse in the case of an inhomogeneous r.f.field, a misset of the pulse flip angle, or off-resonance effects.

3.5 Modifications on the standard experiments

Several modifications on the experiments described in sections 3.2 and 3.3 have been proposed in literature. Most of these modifications are only minor technical differences, leaving the theoretical basis unchanged. In practice these modified experiments are often harder to perform, but they can in certain cases give more accurate or extra information and shorten the execution time. The main modifications proposed so far will be discussed below.

3.5.1 The multiple-echo experiment

In the methods discussed so far, only one response is acquired for each time the sequence is repeated. The sequence has to be repeated many times, for a full set of different t_1-values. Sometimes the same information can be obtained much faster. Consider for example the experiment of Fig.3.12, where only the second half of the last spin echo is detected. It is clear that the information of all other spin echoes, in between the π pulses, gets lost, since these echoes are not acquired. As has been pointed out in ref. (11), detection of the second halves of all spin echoes in a Carr-Purcel-Meiboom-Gill (CPMG) sequence (37) gives a complete two-dimensional datamatrix with t_1-values given by

$$t_1 = 2nT \quad . \qquad n = 0,1,2,\ldots$$

The sampling frequency along the t_1-axis is now given by $(\Delta t_1)^{-1} = (2T)^{-1}$. The maximum sampling time along the t_2-axis is equal to T. In order to avoid folding the sampling frequency along the t_1-axis has to be larger than the multiplet width MW:

$$2T < (MW)^{-1} \quad , \qquad [3.4]$$

which for proton spin systems limits the maximum value for T to a value of the order of 20 msec. The short acqusition time along the t_2-axis implies a poor resolution along the F_2-axis. Fig.3.15a shows the homonuclear J-spectrum of 1,1,2-trichloroethane obtained from a single CPMG-sequence. This experiment is closely related to the one-dimensional J-spectroscopy invented by Freeman and Hill (28), where only one single point at the centre of each spin echo is acquired.

As explained in ref. (38) and (11), acquiring the complete spin echoes makes it possible to generate a two-dimensional absorption spectrum, which partly solves the resolution problem and improves the signal-to-noise ratio. Fig.3.15b shows the absorption J-spectrum of 1,1,2-trichloroethane obtained from the same experiment as the spectrum of Fig. 3.15a, but using the complete echoes. Identical digital filtering was used in the two processings.

There are three major problems connected to this multiple echo method:
(a) The effect of a series of closely spaced π pulses can cause distortion because of apparent spin-locking (39).
(b) The imperfections in the π pulses are not cancelled by the Meiboom-

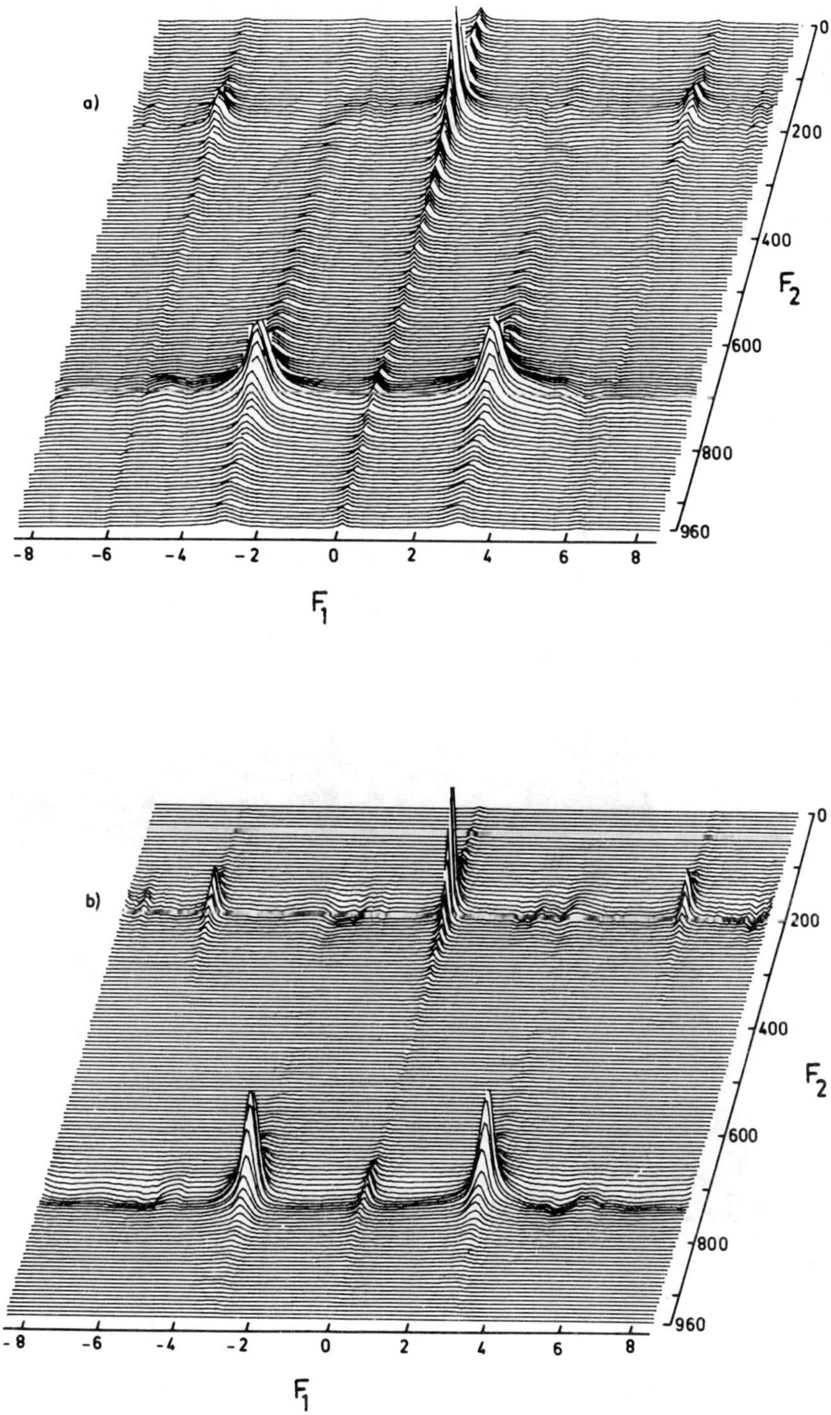

Fig.3.15 Two-dimensional proton J-spectra of 1,1,2-trichloroethane ($CHCl_2CH_2Cl$) at 300 MHz obtained from one single CPMG sequence containing 64 π pulses.
(a) Absolute value mode display. (b) Absorption display. The total measuring time was about 7 sec. From ref.(11).

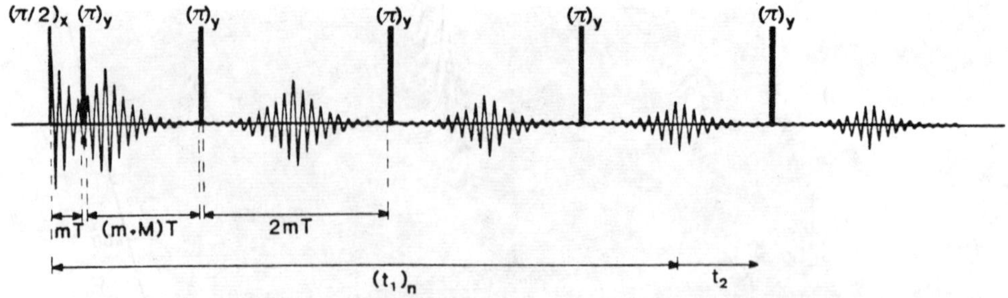

Fig.3.16 Pulse scheme of the m-th spin echo sequence of a relatively shifted multiple echo 2D experiment, used to obtain a J-modulated time signal $M_y(t_1,t_2)$. The complete experiment consists of M similar spin echo sequences. During this sequence a set of second halves of echoes is acquired for $(t_1)_n = 2mT + (n - 1) \times 2T$. Complete echoes can be acquired for $(t_1)_n = 2mT + 2nT$.

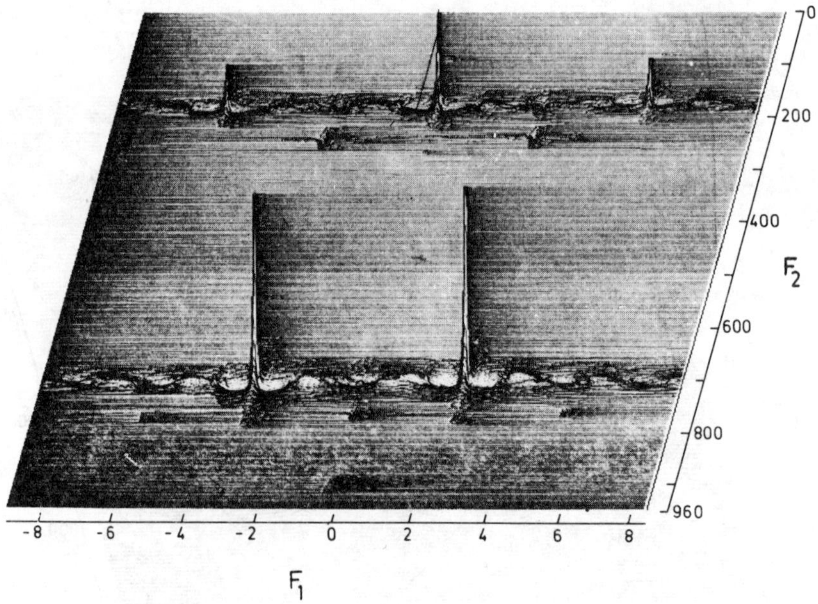

Fig.3.17 2D absorption J-spectrum of 1,1,2-trichloroethane at 300 MHz obtained with the relatively shifted multiple echo experiment. The spectrum was computed from a 512×768 data matrix. The total measuring time was about 10 min. From ref.(11).

Gill modification (40) and have a cumulative effect*.

(c) The resolution along the F_2-axis is still limited because of the short acquisition time along the t_2-axis.

These three problems can partly be solvedby using several (M) CPMG-like sequences (Fig.3.16) in which the centres of the spin echoes are shifted relatively to one another and the spacing between the π pulses can be M times as large (11). This method can be considered as a mixture of the conventional method and the one-shot CPMG method. A spectrum of 1,1,2-trichloroethane, obtained by using eight relatively shifted CPMG-like sequences is shown in Fig.3.17. Note the very high resolution of the multiplet structure (0.04-0.06 Hz). This high resolution is due to the elimination of diffusion effects by the series of π pulses in the evolution period for large values of t_1.

3.5.2 The constant-time experiment

As pointed out in section 3.3 and in ref. (22), the 45° projection of a homonuclear 2D J-spectrum will only generate a decoupled spectrum if performed in the absolute value mode. No phase-sensitive 45° projections with inherent higher resolution can be obtained this way. As pointed out in ref. (10) and (41) there is a way to get a homonuclear broad-band decoupled absorption spectrum. This method, which is also the basis of the experiment discussed in section 2.3.6, will be briefly explained in this section. It is mentioned here that the method described below is strictly not a two-dimensional one, but since it is very closely related to 2D spectroscopy its description is included.

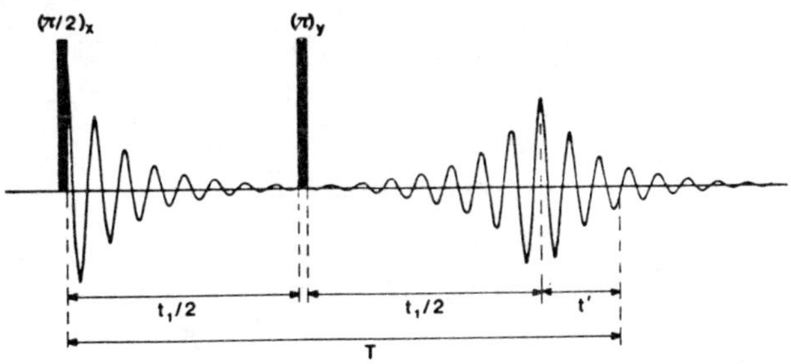

Fig.3.18 Scheme of the method for obtaining homonuclear broadband-decoupled absorption spectra. Sampling points are taken at a fixed time T after the initial $\pi/2$ pulse, for various values of $t_1 \leqslant T$.

---

* This problem can also be solved by using composite π pulses (34, 35, 36).

The pulse sequences of the new method (Fig.3.18) and the standard sequence (Fig.3.1) are quite similar. In the new method a spin echo is created by means of a $(\pi/2)_x - t_1/2 - (\pi)_y - t_1/2$ sequence. A single sampling point at time T after the initial $\pi/2$ pulse is acquired for a set of experiments with different value for t_1. This correspondends to taking a cross-section through the time domain data matrix on the line $t_1 + t_2 = T$. The time domain signal for a resonance line k of a multiplet with chemical shift frequency $\Omega_j/2\pi$ is in the case of weak coupling given by

$$s_{jk}(t') = M_{o\,jk}\exp(-T/T_{2\,jk})\exp(-t'/T_2^{\dagger})\exp(i\Omega_j t' + i\Phi_{jk}) \qquad 0 < t' < T \qquad [3.5]$$

where $t' = T - t_1$, $T_2^{\dagger}$ denotes the decay due to inhomogeneity only and Φ_{jk} is given by

$$\Phi_{jk} = (\Omega_{jk} - \Omega_j)T$$

where Ω_{jk} denotes the angular frequency of component k.
Because of the symmetry in a multiplet of a weakly coupled spin system there will always be a component k' in the same multiplet with opposite phase: $\Phi_{jk'} = -\Phi_{jk}$. The sum of all components k in a multiplet is then given by

$$\sum_k s_{jk}(t') = \sum_k M_{o\,jk}\exp(-T/T_{2\,jk})\exp(-t'/T_2^{\dagger})\cos(\Phi_{jk})\exp(i\Omega_j t') \qquad [3.6]$$

This is a signal with phase zero, rotating with the chemical shift frequency and a strong amplitude dependence on Φ_{jk}, which means on the multiplet splittings and the time T. A Fourier transformation of this signal will give a phase-sensitive broad-band decoupled absorption spectrum. As an example in Fig.3.19 the conventional and the decoupled spectrum of 1,1,2-trichloroethane are shown. To eliminate the amplitude dependence on the term Φ_{jk} and to improve the sensitivity it has been proposed to convert the experiment in an interferogram experiment (41) or to use a two-dimensional version of this experiment (11). A fruitful modification based on this latter principle is the method presented in section 2.3.6.

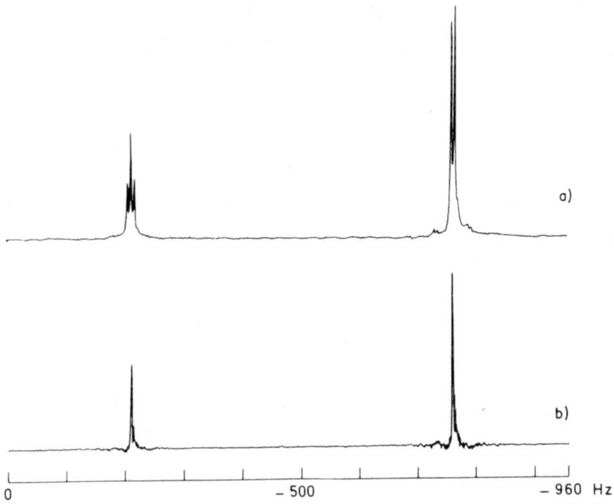

Fig.3.19 ^{1}H spectra of 1,1,2-trichloroethane at 300 MHz (a) obtained with a FID experiment, and (b) the homonuclear broadband-decoupled absorption spectrum. From ref.(38).

3.5.3 Coherent decoupling in homonuclear J-spectroscopy

In order to increase the information contents of a two-dimensional J-spectrum, it has been proposed by Nagayama et al. (9) to perform two separate experiments. The first experiment is the normal standard experiment (Fig.3.1), giving the real splittings in the multiplet. The second experiment only differs from the first one by the fact that continuous coherent irradiation in a certain part in the spectrum is applied. Due to this extra irradiation the multiplet splittings observed in the second 2D spectrum will be reduced by a factor which depends on the offset of the nucleus irradiated from the decoupler frequency. This is similar to the well-known coherent off-resonance proton decoupling in carbon-13 spectroscopy (42). Comparing the multiplets in the two different J-spectra gives an approximate value for the shift of the coupled nucleus. Of course, caution has to be taken in interpreting the results, because of non-linear effects (43) and Bloch-Siegert shifts.

As demonstrated by Nagayama et al. (9) the method can give useful results in proton J-spectra of proteins, where the decoupler is set in the aromatic region of the spectrum, while the aliphatic region is observed.

The opinion of the author is, however, that a combination of a J-spectrum and a homonuclear shift correlation spectrum (section 2.3.2) will give at least the same amount of information, and in a less ambiguous way.

3.5.4 Selective pulses in homonuclear J-spectroscopy

The main problem remaining in homonuclear J-spectroscopy is the lack of resolution. In the standard experiment (section 3.3) this is caused by the phase-twisted line shapes. The multiple echo experiment and the method of section 3.5.2 are not applicable to spin systems with a short T_2, because of the introduced intensity distortions and sensitivity loss.

Nagayama (13) has suggested the use of weak pulses which only affect a part of the spectrum. In this case all kinds of heteronuclear methods can be applied in homonuclear J-spectroscopy, and pure absorption spectra can be obtained. The most practical experiment in the case of using weak pulses is probably the sequence shown in Fig.3.20. Two experiments are performed for each t_1-value. One is the standard pulse sequence, the other one contains a rather weak π pulse to either the observed spectral region or to the region coupled to the region observed. The r.f.field strength of this pulse corresponds typically to a few hundred Hz. To enhance the selectivity of the pulse to a near to perfect π pulse for a part of the spectrum and not affecting any other part, it is strongly recommended to use a composite weak π pulse $\{(\pi/2)_x - (3\pi/2)_y - (\pi/2)_x\}$ (34).

Since the weak π pulse inverts the phase of a multiplet component at time t_1 without changing its frequency, two signals with opposite phase modulation are obtained for each value of t_1. As explained in section 1.5.2 these signals can be combined to yield an amplitude-modulated signal enabling the recording of absorption mode spectra. The details about the processing of the two signals can be found in ref. (12).

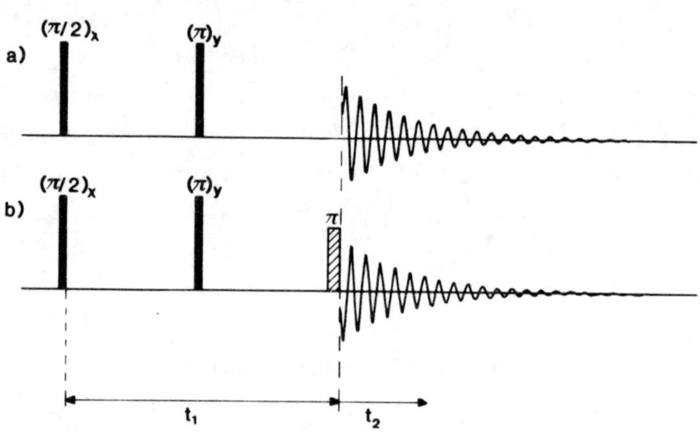

Fig.3.20 Scheme for generation of absorption homonuclear 2D J-spectra. In sequence (b) the weak π pulse at time t_1 can be applied either to the observed spins or to the spins coupled to the observed spins.

Obviously this method can only be used if the nuclei coupled have a large difference in shift. If this is not the case, artefacts in the 2D spectrum will appear.

Demands for flexibility in pulse sequence control and data-processing are very high in this experiment.

References

(1) L. Müller, A. Kumar and R.R. Ernst, J. Chem. Phys. 63, 5490 (1975).

(2) W.P. Aue, J. Karhan and R.R. Ernst, J. Chem. Phys. 65, 4226 (1976).

(3) G. Bodenhausen, R. Freeman and D.L. Turner, J. Chem. Phys. 65, 839 (1976).

(4) G. Bodenhausen, R. Freeman, R. Niedermeyer and D.L. Turner, J. Magn. Reson. 24, 291 (1976).

(5) A. Kumar, W.P. Aue, P. Bachmann, J. Karhan, L. Müller and R.R. Ernst, Proc. XIXth Congress Ampere, Heidelberg 1976.

(6) L. Müller, A. Kumar and R.R. Ernst, J. Magn. Reson. 25, 383 (1977).

(7) G. Bodenhausen, R. Freeman, R. Niedermeyer and D.L. Turner, J. Magn. Reson. 26, 133 (1977).

(8) P. Bachmann, W.P. Aue, L. Müller and R.R. Ernst, J. Magn. Reson. 28, 29 (1977).

(9) K. Nagayama, P. Bachmann, R.R. Ernst and K. Wüthrich, Biochem. Biophys. Res. Commun. 86, 218 (1979).

(10) A. Bax, A.F. Mehlkopf and J. Smidt, J. Magn. Reson. 35, 167 (1979).

(11) A. Bax, A.F. Mehlkopf and J. Smidt, J. Magn. Reson. 40, 213 (1980).

(12) R. Freeman, S.P. Kempsell and M.H. Levitt, J. Magn. Reson. 34, 663 (1979).

(13) K. Nagayama, J. Chem. Phys. 71, 4404 (1979).

(14) L. Müller, J. Magn. Reson. 36, 301 (1979).

(15) E.L. Hahn, Phys. Rev. 80, 580 (1950).

(16) E.L. Hahn and D.E. Maxwell, Phys. Rev. 88, 1070 (1952).

(17) G. Bodenhausen, R. Freeman, G.A. Morris and D.L. Turner, J. Magn. Reson. 28, 17 (1977).

(18) A. Abragam, The Principles of Nuclear Magnetism, p. 308, Oxford University Press, Oxford 1961.

(19) A. Bax, Internal Report, Delft 1978.

(20) G. Bodenhausen and D.L. Turner, J. Magn. Reson. 41, 200 (1980).

(21) R.N. Bracewell, Aust. J. Phys. 9, 198 (1956).

(22) K. Nagayama, P. Bachmann, K. Wüthrich and R.R. Ernst, J. Magn. Reson. 31, 133 (1978).

(23) L.D. Hall and S. Sukumar, J. Magn. Reson. 38, 555 (1980).

(24) K. Nagayama, K. Wüthrich, P. Bachmann and R.R. Ernst, Biochem. Biophys. Res. Commun. 78, 99 (1977).

(25) L.D. Hall, G.A. Morris, S. Sukumar, Carbohydr. Res. 76, C7 (1979).

(26) L.D. Hall, S. Sukumar and G.R. Sullivan, J. Chem. Soc. Chem. Commun. 292 (1979).

(27) A. Bax and R. Freeman, J. Magn. Reson. 37, 177 (1980).

(28) R. Freeman and H.D.W. Hill, J. Chem. Phys. 54, 301 (1971).

(29) C.H.A. Seiter, G.W. Feigenson and S.I.Chan, J. Amer. Chem. Soc. 94, 2535 (1972).

(30) A. Lösche, Kerninduktion, Deutscher Verlag der Wissenschaften, Berlin 1957.

(31) A. Kumar, J. Magn. Reson. 30, 227 (1978).

(32) G. Bodenhausen, R. Freeman, G.A. Morris and D.L. Turner, J. Magn. Reson. 31, 75 (1978).

(33) G. Bodenhausen, R. Freeman and D.L. Turner, J. Magn. Reson. 27, 511 (1977).

(34) R. Freeman, S.P. Kempsell and M.H. Levitt, J. Magn. Reson. 38, 453 (1980).

(35) M.H. Levitt and R. Freeman, J. Magn. Reson. 43, 65 (1981).

(36) R. Freeman and J. Keeler, J. Magn. Reson. 43, 484 (1981).

(37) S. Meiboom and D. Gill, Rev. Sci. Instr. 69, 688 (1958).

(38) A. Bax, A.F. Mehlkopf and J. Smidt, J. Magn. Reson. 35, 373 (1979).

(39) E.J. Wells and H.S. Gutowsky, J. Chem. Phys. 43, 3414 (1965).

(40) R, Freeman and H.D.W. Hill, Dynamic Nuclear Magnetic Resonance Spectroscopy (L.M. Jackman and F.A. Cotton, Eds.), Academic Press, New York 1975.

(41) T.A. Frenkiel, Part II Thesis, Oxford 1980.

(42) R.R. Ernst, J. Chem. Phys. 45, 3845 (1966).

(43) A. Wokaun and R.R. Ernst, Mol. Phys. 36, 317 (1978).

(44) A. Bax, R. Freeman and G.A. Morris, J. Magn. Reson. 43, 333 (1981).

4. MULTIPLE QUANTUM COHERENCE

4.1 Introduction

In this chapter an attempt will be made to explain the phenomenon of multiple quantum coherence in a simple way, and to show its intriguing properties. In the final section its use in high resolution liquid state NMR will be briefly discussed.

In the broadest definition of the term multiple quantum NMR is concerned with the observation of nuclear transitions that are forbidden by the well-known selection rule $\Delta m = \pm 1$. Some of the multiple quantum transitions could already be detected with the old continuous wave method (1,2). In this case the irradiating field is so strong that it significantly disturbs the spin system and modifies its stationary wave functions. Double resonance experiments relying on the same principle, can also be used to detect the forbidden transitions (3,4). The main use of these early multiple quantum expriments was the determination of connectivity and of relative signs of coupling constants.

When Fourier transformation was introduced in NMR there appeared to be no obvious way for it to be used to detect multiple quantum transitions. Detection was considered to be impossible in Fourier transform NMR not only because a single pulse applied to a spin sysytem in thermal equilibrium does not create any multiple quantum coherence, but also because a multiple quantum coherence, possibly present, does not induce any physically observable signal. Almost a decade went by until it was realized that these problems could be circumvented if a more complex excitation sequence is used, and if the detection is done in an indirect way (5-8). The development of Fourier transform multiple quantum NMR has probably been slowed down by the fact that no simple classical picure is available to describe the behaviour of multiple qantum coherence and therefore usually cumbersome density matrix calculations are required (section 1.2, App.I). A calculation method using so-called fictitious spin-$\frac{1}{2}$ operators has been introduced (8,9,10), however, still without giving a really simple view on the physical aspects. If interested, the reader is recommended to study the relatively simple paper by Kaiser on this subject (11).

To give a feeling of what is going on in the new pulsed NMR methods which detect multiple quantum transitions, a theoretically simple experiment will now be discussed qualitatively.

Consider an AX proton spin system as shown in Fig.4.1. First a selective

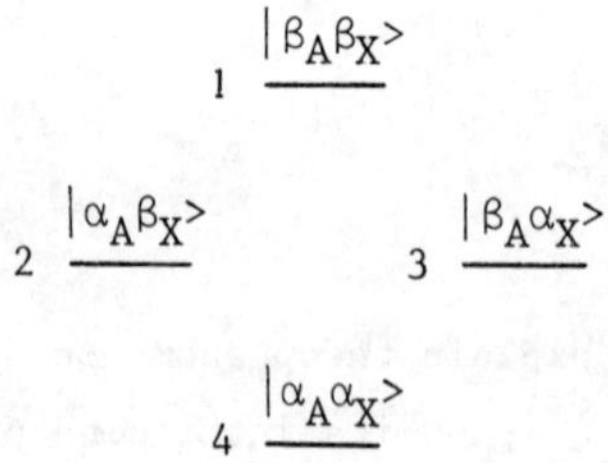

Fig.4.1 Energy level diagram and wave functions of an AX spin system.

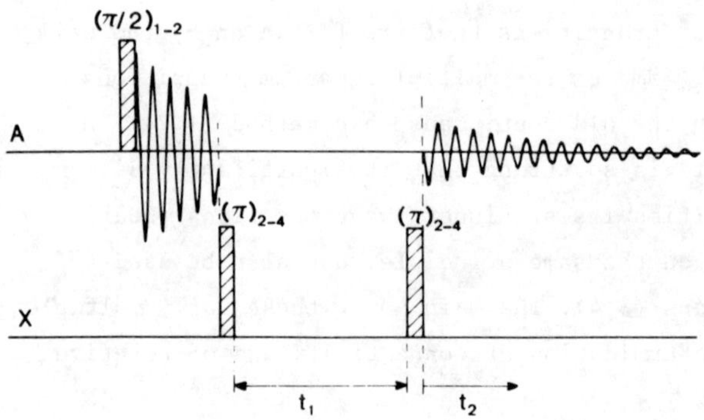

Fig.4.2 Scheme for the selective excitation and detection of the double quantum coherence 1-4 in an AX spin system.

pulse on one of the A-spin transitions (Fig.4.2), e.g. transition 12, creates a coherence between the levels 1 and 2. Then a selective π pulse applied to X-spin transition 24, interchanging the states of levels 2 and 4, transfers the coherence from transition 12 to transition 14. Since transition 14 is a double quantum transition, no signal will be observable after this selective proton pulse. However, the coherence between level 1 and 4 does exist. This can be shown easily by applying another selective π pulse to X-spin transition 24, converting the double quantum coherence back into observable single quantum coherence between the levels 1 and 2, which corresponds to transverse A-spin magnetization. It appears that the phase of the reconverted magnetization is a function of $\omega_{14}t_1$ where ω_{14} is the angular frequency of the double quantum transition and t_1 the time during which the double quantum coherence evolved. By varying the time t_1 in subsequent experiments, the double quantum frequency ω_{14} can thus be determined.

In this chapter only homonuclear multiple quantum transitions will be considered. Heteronuclear multiple quantum transitions rely on the same principles, but involve slightly different experimental aspects (12,13). No important applications for heteronuclear multiple quantum transitions have been proposed so far, which could not be done in a more conventional type of experiment. The importance of homonuclear multiple quantum Fourier NMR is also rather limited (section 4.7). It can be used to obtain extra information about the relaxation mechanism, for spectral simplification, and to separate different spin systems as shown in chapter 5. Furthermore, it is important to understand the phenomenon of multiple quantum NMR, because it can introduce errors in conventional experiments for example relaxation measurements (14,15), even though the multiple quantum frequency is not detected.

4.2 Creation and detection of multiple quantum coherence

As mentioned in the introduction of this chapter, a single non-selective pulse applied to a spin system in thermal equilibrium will create single quantum coherence only (7,16). However, a pulse applied to a spin system which is in a non-equilibrium state will generally create multiple quantum coherence (6,7). Hence, the number of different multiple quantum excitation methods is infinite. Different classes of excitation methods can be distinguished. A subdivision will be made between methods only employing transition-selective pulses, methods employing both transition-selective and non-selective pulses, and methods using non-selective pulses only. In order to detect the effect of multiple quantum coherence it has to be transferred into transverse magnetization by means of either selective or non-selective pulses.

4.2.1 Methods for excitation of multiple quantum coherence

A 1 Cascade of selective pulses

A theoratically simple way to create multiple quantum coherence in a predictable fashion is the use of cascades of selective pulses as shown in Fig. 4.3a (7). Consider e.g. a set of connected single quantum transitions kl, lm and mn. A transition-selective $\pi/2$ pulse creates coherence between e.g. levels k and l. A selective π pulse on transition lm followed by a selective π pulse on transition mn will transfer the originally single quantum coherence kl into coherence between levels k and n. This can easily be verified by using the results given in Appendix III. The advantage of this excitation method is that the amount of multiple quantum coherence is well defined. However, the phase of the coherence can be more difficult to predict since a series of selective pulses at different frequencies has been applied, and the phases of these pulses are thus undetermined. A way around this phase problem can be the use of pulsed selective excitation (17,18), using one transmitter frequency only.

12 *Single selective pulse*

Multiple quantum coherence can be selectively excited if a pulsed r.f.field of appropriate strength and duration is applied at the frequency where the transition considered would show up in the continuous wave method (Fig.4.3b) (5,7,8). The excitation mechanism in this case is the same as in the conventional multiple quantum continuous wave NMR experiments. In liquids this excitation method is mostly not very practical, but it should be realized that a single weak pulse can be capable of generating multiple quantum coherence, which can give rise to undesired effects in all kinds of conventional experiments (15).

31 *Selective π pulse followed by a non-selective π/2 pulse*

As has been shown by Wokaun (7) a convenient way to create all multiple quantum coherences in which a spin A participates, with equal intensity, is to apply a selective π pulse to one of the resonance lines of multiplet A, followed by a non-selective π/2 pulse, applied to the whole spin system (Fig. 4.3c). Instead of a selective π pulse to one of the transitions, a selective

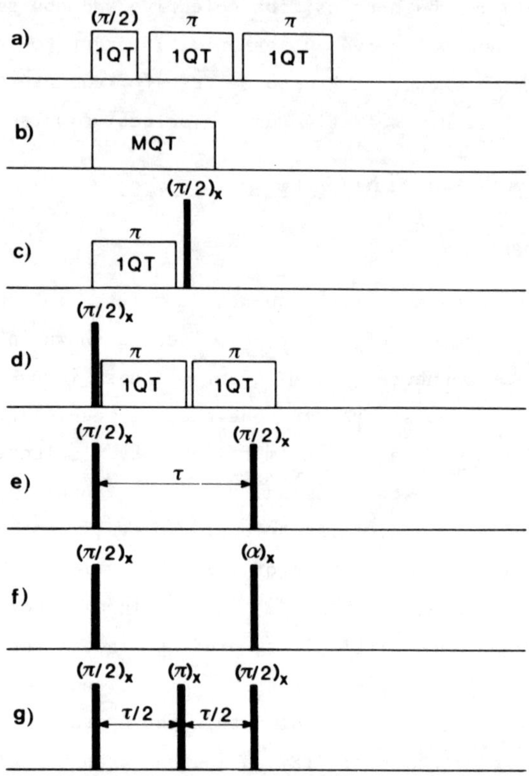

Fig.4.3 Various schemes for the generation of multiple quantum coherence as described in the text.

saturation of this resonance can also be employed, decreasing the amount of multiple quantum coherence created by a factor two. This last method is easily done in practice by using the decoupler irradiation channel for selective saturation. A homospoil pulse has to be applied to eliminate possible spurious transverse magnetization after this saturation. A disadvantage of these methods is that the magnitude of the multiple quantum coherences is rather low compared with the excited single quantum coherences (see e.g. Fig.410a).

B2 Non-selective $\pi/2$ pulse followed by a cascade of selective π pulses

A non-selective $\pi/2$ pulse followed by a cascade of selective π pulses can be used in a similar way to method A1 to create a certain coherence selectively in a predictable fashion (7), and with high intensity (Fig.4.3d). Since the phase of the selective π pulses matters for the phase of the excited coherences, again pulse selective excitation should be used (17,18).

C1 $(\pi/2)_x - \tau - (\pi/2)_x$ excitation

The most common way to excite multiple quantum coherence is the application of two non-selective $(\pi/2)_x$ pulses separated by a delay τ (Fig.4.3e). The transverse magnetization created by the first pulse is redistributed by the second pulse over all possible transitions in the spin system, including the multiple quantum transitions. The major problem using this non-selective excitation is, that without using a density matrix calculation according to Eq.[1.19], the intensities of the transitions are very hard to predict for any coupled spin system consisting of more than two spins. Clearly the multiple quantum excitation efficiency has to depend on the length of the delay time τ. For $\tau = 0$ the two $(\pi/2)_x$ pulses would just have the effect of a single $(\pi)_x$ pulse, not creating any multiple quantum coherence at all. A condition required for the excitation of a double or zero quantum coherence in which two spins A and X are the only spins participating, is that $\sin(\pi J_{AX}\tau)$ is not equal to zero (7). For a two-spin system for example, it follows from an explicit calculation, using Eq.[1.19] that maximum excitation can be achieved if $\sin(\pi J_{AX}\tau)$ equals one.

C2 $(\pi/2)_x - \tau - (\alpha)_x$ excitation

If instead of a $(\pi/2)_x$ an α pulse is used after the delay time τ(Fig.4.3f), where α is smaller than $\pi/2$, mainly zero-, single- and double quantum cocoherence will be excited,i.e. the creation of multiple quantum coherence is more favourable for transitions in which not more than two spins change their polarization. This flip-angle effect has already been discussed in section 2.2.3.

C3 Offset independent excitation

Apart from the value $\sin(\pi J\tau)$ the excitation of multiple quantum coherence using the $\pi/2 - \tau - \pi/2$ method depends on the offset of the centres of multiplets measured with respect to the transmitter frequency. This offset dependence can be removed by placing a π pulse in the centre of the creation period as shown in Fig.4.3g. It can be shown that if the second $\pi/2$ pulse is applied along the same axis as the first pulse, only even orders of coherence will be excited, while if the $\pi/2$ pulses are 90° out of phase $\{(\pi/2)_x - \tau/2 - (\pi)_x - \tau/2 - (\pi/2)_y\}$, only odd orders are excited. This odd - even rule is only valid in the case where the second mixing pulse is a $\pi/2$ pulse. This offset-independent excitation method will be used in chapter 5.

C4 Selective order excitation

A method has been designed by Pines and co-workers which employs non-selective pulses and still generates a selectable order of coherence only, and with high intensity (19,20). This method employs long series of phase-shifted pulses and is both theoretically and practically very complicated.

4.2.2 Detection of multiple quantum coherence

As mentioned before the presence of multiple quantum coherence cannot be detected directly. Its effects can be shown by applying another pulse, as in the example of Fig.4.2. This pulse transfers (part of) the multiple quantum coherence back into observable magnetization. A practical approach for the back transfer is the application of a non-selective $\pi/2$ pulse at time t_1 after the multiple quantum coherence has been created (Fig.4.4). The coherence is then redistributed over all the transitions present in the spin system. It then follows from Eq.[1.30] that all magnetization components after the detection pulse are modulated in amplitude as a function of t_1 by all the frequencies present in the spin system during this evolution period t_1. In principle it is sufficient to detect only a single point at a suitable time after the detection pulse, since this point contains information about the

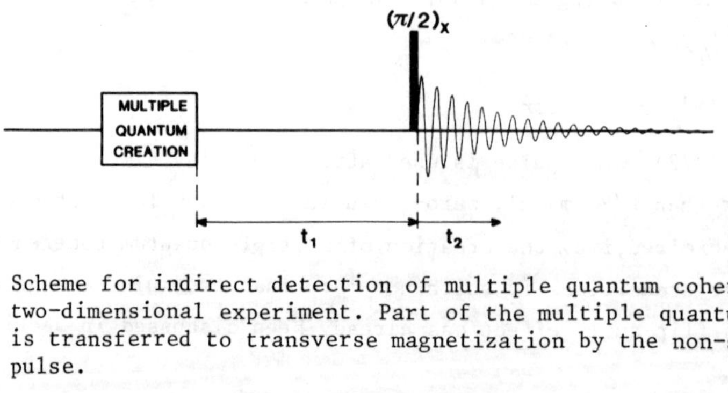

Fig.4.4 Scheme for indirect detection of multiple quantum coherence via a two-dimensional experiment. Part of the multiple quantum coherence is transferred to transverse magnetization by the non-selective $\pi/2$ pulse.

total vector sum of all transverse magnetization components which are all individually modulated with the same frequencies. The modulation frequencies can be determined by a Fourier transformation of the amplitude modulation of this point with respect to t_1(21,22,23). For sensitivity purpose, however, it is better to detect the complete free induction decay for each value t_1. A Fourier transformation of these responses will yield a series of spectra in which all the individual lines are modulated in amplitude as a function of t_1 with all the single and multiple quantum frequencies present in the spin system. A Fourier transformation of this data matrix with respect to t_1 will now give a two-dimensional spectrum with all the multiple quantum frequencies along the F_1-axis. A projection of such a spectrum in the absolute value mode on the F_1-axis will produce a one-dimensional spectrum, showing all the multiple quantum frequencies which were present in the spin system. As an example a two-dimensional spectrum of acrylonitrile and its projection obtained this way, using excitation method of Fig.4.3e, is shown in Fig.4.5.

Single point acquisition

As has been shown in section 1.2.4, a non-selective pulse cannot transfer multiple quantum coherence into net transverse magnetization, i.e. the vector sum of the magnetization components, originating from multiple quantum coherence is zero, immediately after this pulse. Hence, in order to observe any signal transferred from multiple quantum coherence a delay τ' after the detection pulse is required before magnetization is sampled. This is the procedure actually used in the case where only one point is sampled for each value of t_1 as is for instance the case in the multiple quantum studies in liquid crystals as performed by Pines and co-workers (22,23). As they use a $\pi/2 - \tau - \pi/2$ multiple quantum excitation sequence, the value τ' is chosen to be equal to τ.

To eliminate intensity distortions in the multiple quantum spectrum due to the fact that one value for τ will generally not be equally efficient for the observation of all different multiple quantum transitions, several experiments are performed for different values of τ, and the resulting absolute value mode multiple quantum spectra are co-added.

Multiple point acquisition

The procedure described above can in principle also be used in multiple quantum studies in liquids, but is inefficient and time-consuming. An averageing over different values of the delay τ' is automatically obtained if a two-dimensional experiment is performed where the signal after the detection

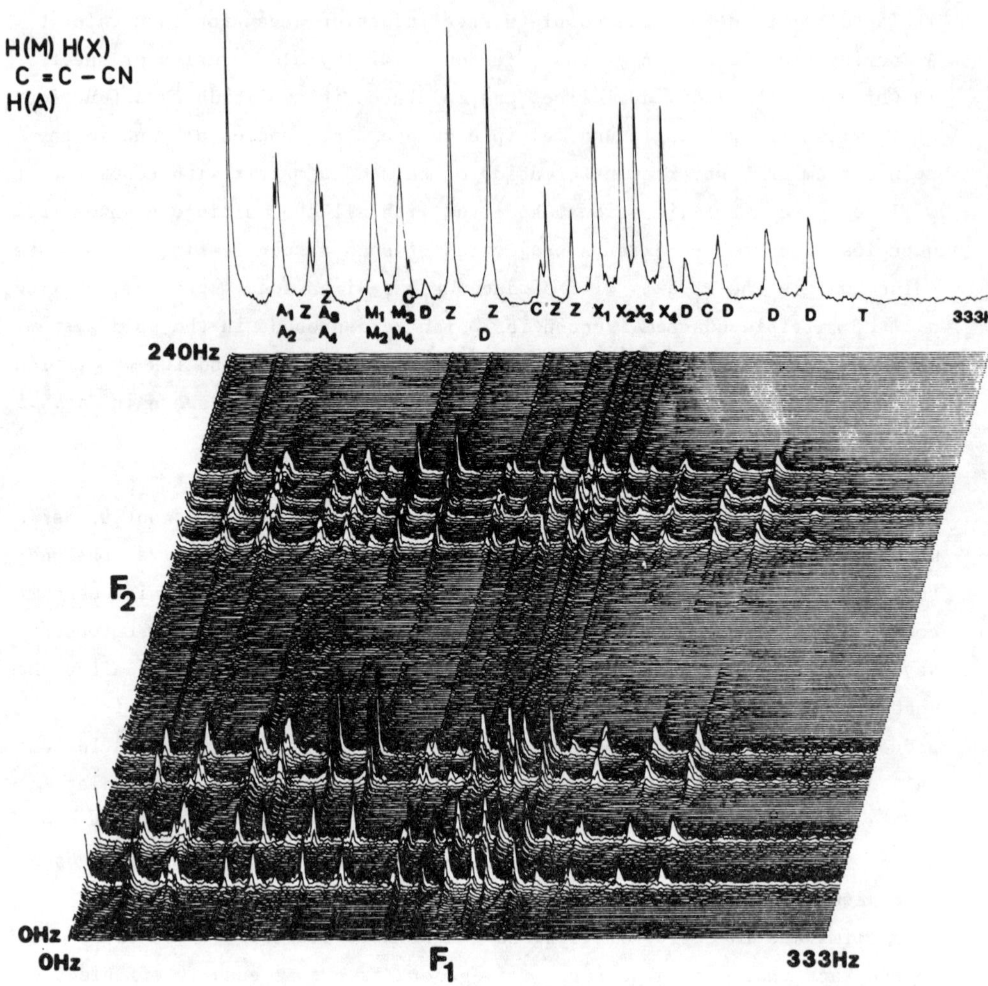

Fig. 4.5 Two-dimensional multiple quantum spectrum of the proton AMX system of acrylonitrile and its projection onto the F_1-axis. Under the projection the single quantum resonance lines are labelled A_1-A_4 for spin A etc., The combination lines are indicated by C, the zero quantum resonance lines by Z, the double quantum resonance lines by D and the triple quantum resonance by T. The spectrum is recorded at 300 MHz.

pulse is monitored for a time t_2. Intensity anomalies now will only depend on the choice of the delay τ in the excitation scheme, and will be significantly less severe. However, a large data matrix is generally needed for data storage in this case. Since a non-selective pulse does not create net magnetization and magnetization components have opposite phases at $t_2 = 0$, the intensities of resonance lines in the two-dimensional phase-sensitive multiple quantum spectrum will either be positive or negative, in such a way that the total intensity integral over the spectrum is equal to zero. For this reason the absolute value mode is commonly used in the display of multiple quantum spectra.

Since a non-selective pulse creates amplitude modulation of the detected signal with respect to t_1, the positive and negative multiple quantum frequencies cannot be distinguished (section 1.5.2). A conversion of this amplitude modulation into phase modulation analogous to the methods described in sections 2.2.3 and 2.3.2 is in general not possible.

4.2.3 Frequency of multiple quantum coherence

One might expect that a p quantum transition has a frequency in the order of p times the Larmor frequency. In principle this is true, but it has to be realized that the experiment is actually performed in a frame, rotating with Ω_R, where Ω_R is the transmitter frequency which is equal to the reference frequency in the lock in detector. The transformation to a rotating frame has the effect of decreasing the multiple quantum frequency expected by $p\Omega_R$ (Eq. [1.13]). This gives the convenient result that the multiple quantum frequencies detected will be equal to a sum or difference of resonance offset frequencies as observed in the corresponding single quantum spectrum. For example in an AX spin system the double quantum frequency ω_{14} will be observed at the sum of the offset frequencies ω_{12} and ω_{24} with respect to the transmitter frequency. In a similar way the zero quantum resonance frequency ω_{23} will be equal to the difference of single quantum frequencies $\omega_{12} - \omega_{13}$.

4.2.4 Which multiple quantum coherence can be observed?

There appear to be restrictions with respect to which multiple quantum coherences can be created. It appears not to be true that any connection between two energy levels in a spin system represents a possibly detectable multiple quantum coherence. In this section the fact will be used that if it is impossible to create a certain coherence by means of a cascade of selective pulses, it will also be impossible by means of non-selective pulses, since the latter can always be considered as a cascade of selective pulses (App.III, Eq. [III.20]).

Symmetry group restrictions

In the case of magnetically equivalent nuclei the symmetrized stationary wave functions are divided into symmetry groups (24). Only multiple quantum coherences within a symmetry group can be created. Consider e.g. the simplest case of an AX_2 spin system (Fig.4.6). Only multiple quantum coherences in the symmetric part of the energy level diagram are possible, since it is impossible to apply a selective pulse which transfers magnetization between different symmetry groups.

$(X)_n$-transitions

In the case of isolated coupled magnetically equivalent nuclei which do not have scalar interaction with other nuclei in the same molecule, as e.g. the protons in benzene, it is impossible to create a multiple quantum coherence. However, if magnetically equivalent nuclei are coupled to another nucleus and form e.g. an AX_2 or AX_3 spin system, a multiple quantum coherence can be created in which only the equivalent nuclei participate, provided that both the A and X nuclei are affected in "multiple quantum excitation sequence."

Vanishingly small coupling

In the case where coupling between nuclei A and X is vanishingly small it will be impossible to create a multiple quantum coherence in which only spins A and X participate. Again this can be understood by the fact that it is impossible to apply a transition-selective pulse to one of the A-transitions only, because of the overlap of the two A-multiplet components. How-

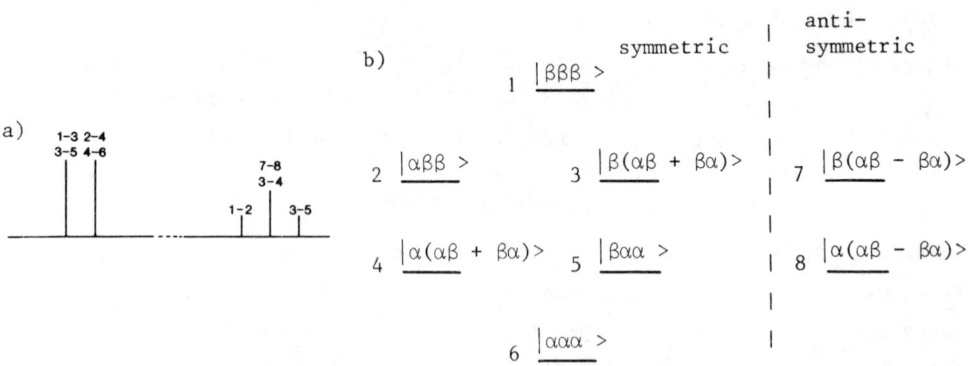

Fig.4.6 The conventional spectrum (a) and the energy level diagram (b) for an AX_2 spin system.

ever, if another nucleus M exists, which is coupled to both the A- and the X-nucleus, it is possible to create a triple quantum coherence in which all three spins A, M, and X participate.

4.3 Special properties of multiple quantum coherence

Multiple quantum coherence appears to have a sensitivity to magnetic field strength(or transmitter offset frequency) and phases of r.f.pulses which is characteristic for their order (6,25). Both effects have the same theoretical basis. The effect of an extra offset $\Delta\omega$ on the frequency of coherence of order p is a change by $p\Delta\omega$ (Eq. [1.13]). The effect of a phase shift ϕ of the excitation pulses, on coherence of order p is that it causes a change in phase of this coherence by $p\phi$. Both effects are incorporated in the formula

$$\exp(i\phi\hat{F}_z)\sigma_p\exp(-i\phi\hat{F}_z) = \exp(ip\phi)\sigma_p \qquad [4.1]$$

where σ_p represents a density matrix only containing p-quantum elements σ_{kl} with $M_k - M_l = p$ and the normal density matrix $\sigma = \sum_{p=-N}^{N} \sigma_p$, where N is the number of spins in the system. More explicitly the effect on a matrix element $\sigma_{kl}(t_1)$ describing coherence between state k and l at time t_1 can be written

$$\exp(i\phi\hat{F}_z)\sigma_{kl}(t_1)\exp(-i\phi\hat{F}_z) = \sigma(0)\exp\{i\phi(M_k - M_l)\}\exp(i\omega_{kl}t_1), \qquad [4.2]$$

where $\sigma(0)$ is the density matrix at time $t_1 - 0$. The operator $\exp(i\phi\hat{F}_z)$ denotes a rotation over an angle ϕ about the z-axis. This rotation can either be caused by (a) the static magnetic field along the z-axis, or (b) a phase change of the radiofrequency pulses.

ad a. If the static magnetic field is described by a perfectly homogeneous component H_o and an inhomogeneity contribution $H_o(r)$ the evolution of the density matrix is described as

$$\sigma(t_1) = \exp\{-i\gamma H_o(r)\hat{F}_z t_1\}\exp(-i\gamma H_o\hat{F}_z t_1)\sigma(0)\exp(i\gamma H_o\hat{F}_z t_1) \times \qquad [4.3]$$
$$\times \exp\{i\gamma H_o(r)\hat{F}_z t_1)\}.$$

With ϕ equal to $-\gamma H_o(r)t_1$, Equations [4.3] and [4.1] show directly the relation that the dephasing of coherence due to inhomogeneity is proportional to the order of the coherence. This agrees with the earlier statement that multiple quantum frequencies are always equal to sums or differences of single quantum transition frequencies (section 4.2.3). Consider for example the AX spectrum in Fig.4.7 in which lines are broadened by an inhomogeneous magnetic field. The shaded and the black areas of the resonance lines cor-

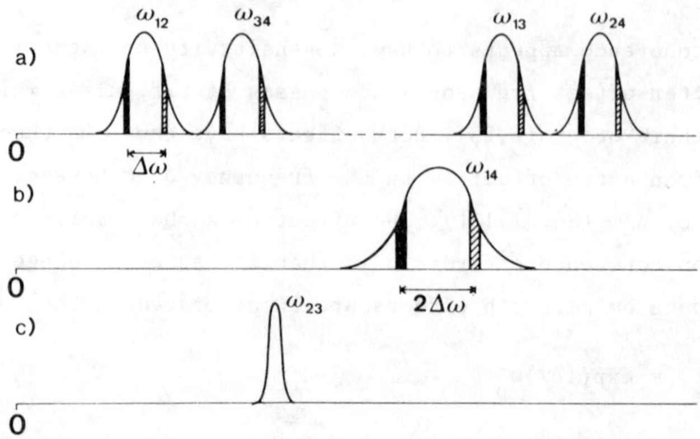

Fig.4.7 (a) Schematic single quantum spectrum of an AX spin system (Fig.4.1) in an inhomogeneous magnetic field. Black and shaded areas correspond to two different regions in the sample.
(b) The double quantum spectrum.
(c) The zero quantum spectrum.

respond to two positions in the sample differing an amount $\Delta\omega$ in frequency. Since the double quantum frequency is equal to the sum of ω_{13} and ω_{34} or ω_{12} and ω_{24}, the difference in double quantum frequency between those two regions is $2\Delta\omega$. The zero quantum transition, at a frequency equal to the difference between ω_{12} and ω_{13} will clearly be independent of inhomogeneity broadening.

<u>ad b.</u> The rotation operator corresponding to a pulse with flip angle α applied along an axis in the xy-plane which makes an angle ϕ with the positive x-axis is given by

$$\hat{R}_\phi(\alpha) = \exp[i\alpha\{\hat{F}_x\cos(\phi) + \hat{F}_y\sin(\phi)\}] = \exp(-i\phi\hat{F}_z)\exp(i\alpha\hat{F}_x)\exp(i\phi\hat{F}_z). \quad [4.4]$$

If for example in the $(\pi/2)_x - \tau - (\pi/2)_x$ multiple quantum excitation sequence the $(\pi/2)_x$ pulses are replaced by $(\pi/2)_\phi$ pulses this has the following effect:

$$\begin{aligned}\sigma_\phi(0) &= \hat{R}_\phi(\pi/2)\exp(-i\hat{H}\tau)\hat{R}_\phi(\pi/2)\sigma_o\hat{R}_\phi(-\pi/2)\exp(i\hat{H}\tau)\hat{R}_\phi(-\pi/2) \quad [4.5]\\ &= \exp(-i\phi\hat{F}_z)\hat{R}_x(\pi/2)\exp(-i\hat{H}\tau)\hat{R}_x(\pi/2)\sigma_o\hat{R}_x(-\pi/2)\exp(i\hat{H}\tau)\hat{R}_x(-\pi/2)\exp(i\phi\hat{F}_z)\\ &= \exp(-i\phi\hat{F}_z)\sigma_x(0)\exp(i\phi\hat{F}_z),\end{aligned}$$

where σ_o is the equilibrium density matrix and using the facts that $\hat{H}$ and

$\hat{F}_z$ commute and σ_o does not contain any off-diagonal elements. $\sigma_\phi(0)$ is the multiple quantum density matrix at time $t_1 = 0$ in the case where the excitation pulses are applied along the ϕ-axis, while $\sigma_x(0)$ describes the system at $t_1 = 0$ in the case where both pulses were applied along the x-axis. From Eqs. [4.5] and [4.2] it is clear that a change in phase ϕ of the excitation pulses causes a change $-p\phi$ in the phase of a coherence between the two states which differ an amount p in magnetic quantum number. In a similar way it can be proved that a phase change ϕ in the detection pulse has the same effect as an increase in the phase of the multiple quantum coherence of $(p + 1)\phi$ which can also be understood by the fact that only the relative phases of excitation and detection pulses matter.

4.4 Separation of different orders of multiple quantum coherence

As can be seen in Fig.4.5 the multiple quantum spectrum is crowded and looks rather complicated, even for a simple AMX spin system. The main reason for this is the large number of resonance lines due to the fact that resonance lines originating from all orders appear in the same spectrum and show extensive overlap. There are different ways to remove this overlap, i.e. separate the different orders.

4.4.1 Resonance offset frequency

The simplest way to get the orders separated is to set the transmitter frequency at a large enough frequency difference $\Delta\omega$ away from the centre of the one-dimensional spectrum. The effect of this offset $\Delta\omega$ is to move resonance lines in the multiple quantum spectrum over a distance $p\Delta\omega$ away from zero frequency. Since multiple quantum spectra can be considerably wider than the single quantum spectrum, the offset has often to be much larger than the single quantum spectral width.

If a full two-dimensional spectrum is recorded this implies a large increase in the size of the data matrix, because the sampling frequency along both time axes has to be increased. In the case where only a single data point is acquired for each value of t_1, the consequence is that the bandwidth of the audiofrequency filter has to be set to a large value in order to detect the single quantum transverse magnetization. In principle the consequent loss in sensitivity can be eliminated if a band-pass audiofrequency filter is employed which attenuates all frequencies except the band in which the single quantum spectrum is to be found.

4.4.2 Pulsed field gradients

Another simple method to separate the different orders is based on the different sensitivity to magnetic field inhomogneity and is described in detail in ref.(26). The idea is to cause a certain amount of dephasing of the multiple quantum coherence in different parts of the sample, by applying a pulsed magnetic field gradient FG during a time T. After the multiple quantum coherence of a certain order - still present but out of phase - is reconverted into transverse magnetization, the magnetization originating from defocussed p-quantum coherence can be refocussed by the application of another pulsed field gradient of the same intensity but p times as long (Fig.4.8). With this method only one order can be selected per experiment, unless several individually spaced field gradient pulses with length T are applied during the detection period, with sampling of the signals in between (26). Since the detected signal after a single non-selective $\pi/2$ pulse is amplitude modulated by the multiple quantum frequencies, there will be a coherence transfer echo and an antiecho (25) (section 1.3). Both these signals are phase modulated, but by frequencies of opposite sign. If the pulsed field gradients in Fig.4.8 have the same polarity, the coherence transfer echo will be selected, in the case of opposite polarity the antiecho will be selected. Because the detected signal is now phase modulated, positive and negative multiple quantum frequencies can be distinguished, i.e. the transmitter can be set to the centre of the spectrum to minimize the size of the data matrix required without introducing folding. An example of the order-separated spectra of acrylonitrile is shown in Fig.4.9. The minimum time to obtain a multiple quantum spectrum of a certain order, using this separation method, equals the required number of increments for t_1 times the time each experiment takes. The spectra shown in Fig.4.9 were obtained from 512 × 192 data matrices, taking an acquisition time of 512 × 6 seconds (≃1 hour) each.

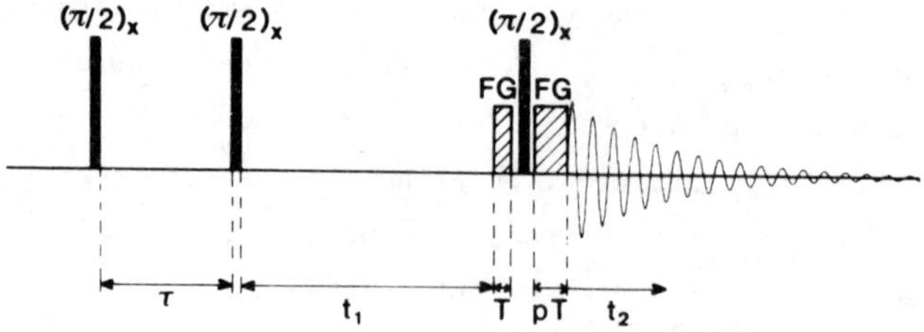

Fig.4.8 Scheme for the selective detection of multiple quantum coherence of order p, using pulsed field gradients (FG).

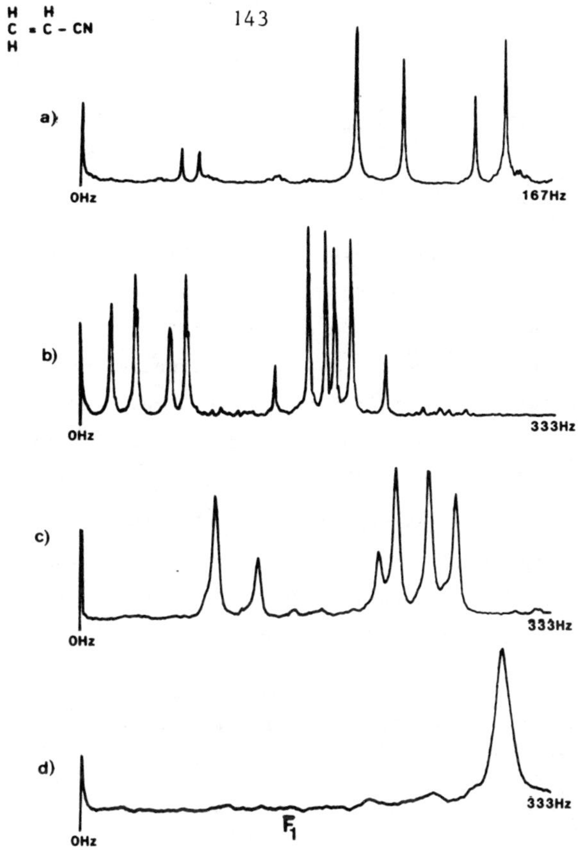

Fig.4.9 Separated order multiple quantum spectra of acrylonitrile, an AMX proton spin system at 300 MHz, in a rather inhomogeneous magnetic field.(a)The zero quantum spectrum, (b) the single quantum spectrum including the combination lines, (c) the double quantum spectrum and (d) the triple quantum spectrum. The four different spectra are obtained from four different two-dimensional experiments, using pulsed field gradients for the selection of the different orders. Difference in line width for the different orders is clearly visible.

4.4.3 Separation based on phase properties

As is shown in section 4.3 a change ϕ in the phase of all the excitation pulses causes an apparent rotation around the z-axis which means an increment $p\phi$ in phase of the multiple quantum coherence of order p. Experimentally slightly different tricks to separate the orders, all relying on this principle, are published by the groups of Ernst (13), Pines (22), Vold (27) and Freeman (28,29). Wokaun and Ernst showed that if for each value of t_1 a series of experiments was performed with different phases of the excitation (or detection) pulses, a proper linear combination of the different responses obtained for each value of t_1, can be used to select a certain order.

The transverse magnetization detected along the x- or y-axis, in the case where the excitation pulses are applied along an axis which makes an angle ϕ

with the x-axis is according to Eq. [4.5] given by

$$M_{x,y}(t_1,t_2,\phi) = \sum_{p=-N}^{N} \mathrm{Tr}\{\hat{F}_{x,y}\sigma_{px}(t_1,t_2)\}\exp(-ip\phi) \ , \qquad [4.6]$$

where the index x in $\sigma_{px}(t_1,t_2)$ denotes that this is the matrix $\sigma_p(t_1,t_2)$ if the excitation pulses were applied along the x-axis. Magnetization, originating from coherence of order p, can now be selected by performing several complete two-dimensional experiments for different values of ϕ, and calculation of a proper linear combination of the detected signals. As an example, odd and even quantum coherences can be separated by performing two experiments, one with $\phi = 0$ and one with $\phi = \pi$. From Eq. [4.6] follows, that if the sum of both results is taken, only signals from even quantum coherence will remain, while the difference will give signals originating from odd quantum coherence. If only four different phases for the r.f.pulses are available, the orders which can be separated are given In Table 5.1.

Table 5.1 Simple combinations of results of experiments with phase-shifted excitation to observe certain orders of coherence selectively.

Phase ϕ	0°	90°	180°	270°	Observed orders
Adding or Subtracting	+				0 1 2 3 4 5 6 7 8 9
	+		+		0 2 4 6 8
	+		−		1 3 5 7 9
	+	+	+	+	0 4 8
	+	−	+	−	2 6

As explained by Wokaun (14) expression [4.6] can be Fourier transformed with respect to ϕ, selecting the different orders p. This Fourier transform is actually the same as taking linear combinations of responses obtained with different values ϕ.

If the combination of the different signals is done on-line, only a relatively small data matrix is required. However, if the combination of signals is calculated afterwards, an n times larger amount of data storage capacity is required, if n is the number of different experiments for each t_1 value. This latter approach does have the advantage that all different order spectra can be extracted from one set of data. If the combination of data is done in such a way that only one value of p in Eq. [4.6] is selected, either the coherence transfer echo or the antiecho is selected, depending on whether p is negative or positive. Since these signals are phase-modulated, positive and negative multiple quantum frequencies can be distinguished (section 1.5.1). In the combination method of discriminating zero and single quantum signals, shown above, both positive and negative values of p are selected simultaneously, leading to amplitude modulation and making it impossible to distinguish between positive and negative modulation frequencies (section 1.5.2). The phase cycling proposed by Wokaun in ref.(7) gives only amplitude modulated signals. In principle N + 1 different experiments for each value of t_1 are sufficient to separate all N + 1 orders in a N spin system, while 2N + 1 experiments are needed in order to detect the echo or the antiecho of a certain coherence selectively.

The practical approach proposed in ref.(22,27,28,29) and generally referred to as time proportional phase increment (TPPI) is slightly different. Only one value for the phase ϕ is used per value of t_1. However, each time t_1 is incremented the value of ϕ is also incremented by $\Delta\phi$. This causes according to Eq. [4.5] an extra rotation $p\Delta\phi$ around the z-axis per time unit Δt_1. This means that the multiple quantum frequency of order p is apparently increased by $p\Delta\phi/\Delta t_1$. Since the normal multiple quantum modulation frequency ω_p, increased by $p\Delta\phi/\Delta t_1$ is sampled at discrete intervals, this implies for the sampling frequency $(\Delta t_1)^{-1}$ in order to avoid folding:

$$(\Delta t_1)^{-1} > (\omega_p + \frac{N\Delta\phi}{\Delta t_1})/\pi . \qquad [4.7]$$

This means that $\Delta\phi$ has to be smaller than π/N. Usually $\Delta\phi$ is chosen to be equal to

$$\Delta\phi = \pi/(N + \tfrac{1}{2}) \qquad [4.8]$$

if N is the highest order of transitions to be detected. Since all multiple quantum frequencies of order p now have an offset of $p(\Delta t_1)^{-1}/(2N + 1)$, positive and negative multiple quantum frequencies are automatically discriminated. In order to produce phase shifts which are not an integer multiple of $\pi/2$, and are not available on standard NMR spectrometers a digital phase shifter for this purpose can be built, as suggested by Bodenhausen (30). An elegant alternative circumventing the extension of hardware is the use of a composite z-pulse (28,29). In this case the normal excitation pulses are left unchanged in the successive experiments but a ϕ_z composite pulse $\{(\pi/2)_x - (\phi)_y - (\pi/2)_{-x}\}$ is applied during the evolution period. The angle ϕ is now incremented by $\Delta\phi$ each time t_1 is incremented. The theoretical description of this experiment is the same as in the TPPI method. As the sampling frequency along the t_1-axis has now to be 2N + 1 times as high as in the normal experiment, a 2N + 1 times larger data matrix is required. This is the same as the amount needed if the combination method described earlier and distinguishing positive and negative signals is used. Using the methods described in this section, the number of experiments to be executed is 2N + 1 times larger than to obtain a non-separated spectrum, thus lengthening the minimum experiment execution time by a factor 2N + 1.

Experimental

Multiple quantum spectra of acrylic acid obtained by using the TPPI method with composite z-pulses are shown in Fig.4.10. The phase increment $\Delta\phi$ used was $\pi/3$. 512 increments for t_1 were used and 256 complex data points along the t_2-axis were sampled. The sampling frequency along the t_1-axis $(\Delta t_1)^{-1}$ was 1500 Hz. Spectrum b) was obtained using a $(\pi/2)_x$ - 40 msec. - $(\pi/2)_x$ excitation, while spectrum a) was obtained by using excitation method of Fig.4.3c (section 4.2.1), selectively inverting a resonance line in the multiplet of spin M. In spectrum a) all multiple quantum resonances in which

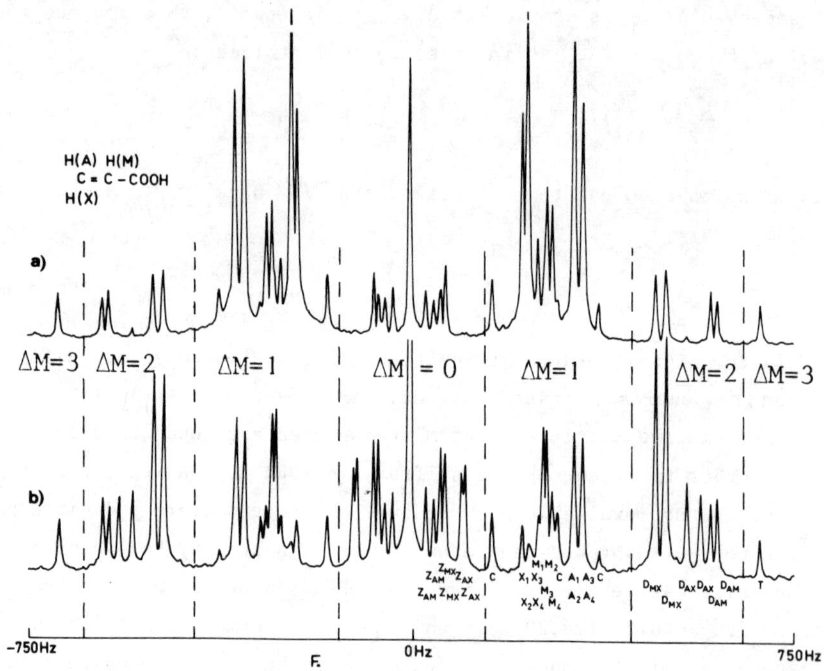

Fig.4.10 Order-separated multiple quantum spectrum of acrylic acid at 200 MHz, obtained by using the TPPI method for the separation, (a) using a line-selective π, non-selective $\pi/2$ excitation and (b) using a $(\pi/2)_x$ - 40 msec. - $(\pi/2)_x$ excitation. The left halves of the two spectra correspond to the antiecho part, the right halves to the coherence transfer echo part. The triple quantum resonances are folded. The spectra are recorded by T.A. Frenkiel.

this M spin participates show the same intensity, as predicted in ref.(14) and in section 4.2.1. Both spectra are symmetrical around the centre, due to the fact that the detected signal is still modulated in amplitude rather than in phase, and a complex Fourier transform with respect to t_1 is used. The right half of the spectra corresponds to the coherence transfer echo, while the left half corresponds to the antiecho. Since the acquisition time along the t_1-axis was only 340 msec. strong Gaussian weighting was applied using a time constant $T_G = 0.2$ sec., which causes so much broadening that no clear difference in linewidth between the different orders can be observed.

4.5 Measurement of relaxation rates

An important application of multiple quantum transitions in liquids is the independent and sometimes extra information they give about the transverse relaxation mechanism (section 4.7) since degeneration of transitions can be removed, and different spectral densities contribute to the relaxation (31,32) Therefore in this section attention will be given to the experimental aspects of multiple quantum relaxation measurements. Since the contribution of magnetic field inhomogeneity to the decay rate can be quite significant, especially for higher orders, a way has to be found to eliminate inhomogeneity broadening.

Zero quantum coherence

As mentioned in section 4.3, zero quantum coherence is insensitive to magnetic field inhomogeneity (14,25,26) and in the case of single exponential decay its relaxation rate can be directly determined from the line width using the well known formula $\Delta\nu_{0.5} = (\pi T_2)^{-1}$. Care has to be taken because this formula only holds for absorption mode spectra. In the absolute value mode presentation, lines will be $\sqrt{3}$ broader. Of course, the signal should be sampled for a sufficiently long time along the t_1-axis: $(t_1)_{max} \geqslant 3T_2$, in order to avoid line shape distortion. Zero-filling by a factor of 2 to 5 before Fourier transformation is recommended to yield proper digitization of the zero quantum line shape.

Non-zero order coherence

If the relaxation rate of a non-zero order coherence is to be determined a different approach is required. As has been shown by Wokaun (7,32), the amplitude of the coherence transfer echo at time pt_1 after the mixing pulse is in principle not influenced by magnetic field inhomogeneity. The amplitude of the coherence transfer echo due to transfer from multiple quantum coherence mn (order p) to single quantum coherence kl decays by the product $\exp(-t_1/T_{2_{mn}})\exp(-pt_1/T_{2_{kl}})$. As the transverse relaxation time $T_{2_{kl}}$ can be determined in a conventional experiment, it is in principle possible to measure $T_{2_{mn}}$ this way. However, this method will not be very accurate since $T_{2_{mn}}$ is determined from the difference of two decay rates which both have experimental errors.

A more practical approach has been proposed by Bodenhausen et al. (27). A π pulse is used in the centre of the evolution period, refocussing the inhomogeneity decay (4.11).Relaxation rates can now be immediately be determined, via line width determination. An important extra advantage of this trick is that in the case of weak coupling the modulation frequencies are only determined by scalar coupling(analoguous to homonuclear J-spectroscopy described in section 3.3), yielding a so-called multiple quantum J-spectrum. This allows the use of a low sampling frequency along the t_1-axis leading to an acceptable size of the data matrix required. However, two disadvantages are connected to this approach:

1. All orders will be overlapping since line positions are not influenced by chemical shifts. This problem can be circumvented by using e.g. the TPPI method (22). Overlap within one order can then still occur if for example two different coupling constants in the molecule have similar magnitudes. Since the effects of offsets are refocussed, orders cannot be separated by using a large offset frequency (section 4.4.1).
2. For all orders except the highest, the relaxation rate determined will be the average of the two relaxation rates, since the π pulse transfers the multiple quantum coherence to its counter multiplet partner. This is the same effect as occurs in conventional transverse relaxation measurements in a homonuclear coupled spin system, using a $\pi/2 - \tau - \pi - \tau$-acquire se-

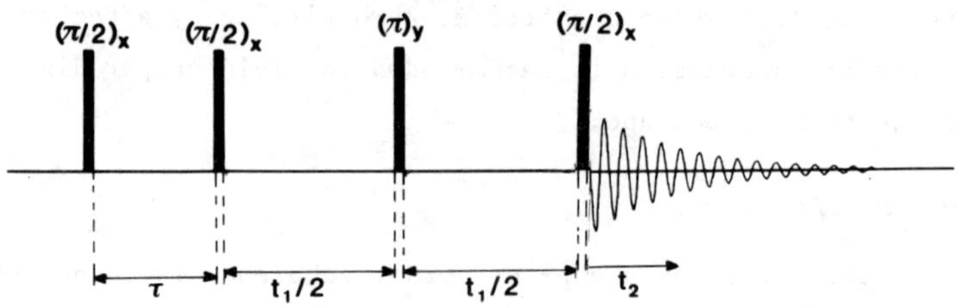

Fig.4.11 Scheme for multiple quantum J-spectroscopy. The π pulse in the centre of the evolution period refocusses the effects of static magnetic field inhomogeneity.

quence (33).

Concluding it can be stated that multiple quantum relaxation rates can be obtained most conveniently by line width determination in a multiple quantum J-spectrum. Zero quantum relaxation rates might be obtained from the normal zero order spectrum, where either a field gradient pulse during the evolution period or the phase cycling as proposed by Wokaun (13)(section 4.4) can be used to suppress the other orders. In this case the relaxation rates of zero quantum multiplet components are detected individually.

4.6 A fast method for the detection of zero quantum coherence

Zero quantum coherence is a particularly interesting class of multiple quantum coherence, since zero quantum transitions were not accessible by the early continuous wave methods and because their insensitivity to magnetic field inhomogeneity allows an accurate determination of the relaxation rate. Because of the insensitivity to field inhomogeneity it is possible to design a sequence which makes it possible to detect zero quantum transitions in a pseudo one-dimensional single shot experiment, as proposed in ref.(34). The idea is based on the fact that in a strongly inhomogeneous magnetic field all coherences except the zero order will defocus rapidly. By applying small pulses the evolution of the zero quantum coherence can be monitored.

The pulse scheme actually used for this experiment is set out in Fig.4.12.

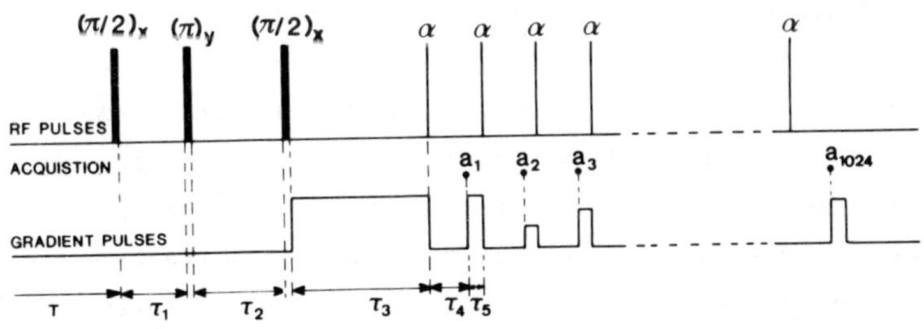

Fig.4.12 Pulse sequence used to excite and detect zero quantum frequencies. Transverse nuclear magnetization was sampled at the points indicated by a_1, a_2,..., a_{1024}; immediately after each sample, the transverse magnetization was dispersed by a gradient pulse of random amplitude. The lengths of the various delays in msec. are: T = 3000, τ_1= 14, τ_2= 18, τ_3= 20, τ_4= 5, τ_5= 1.

In principle all excitation schemes which generate zero quantum coherence can be used. In Fig.4.12 basically a $\pi/2 - \tau - \pi/2$ excitation is used. The π pulse slightly offset from the midpoint of the interval between the two $\pi/2$ pulses has the effect that it partly refocusses the effect of field inhomogeneity during this period $\tau_1 + \tau_2$. During the next interval a strong magnetic field gradient is applied along the z-axis, causing a rapid decay of all non-zero order coherences. This gradient corresponded to a line width of approximately 3 kHz. In addition the room temperature field correction coils were shut off for the entire experiment and no sample spinning was used. This gave an instrumental line width of the order of 150 Hz, at a Larmor frequency of 300 MHz. The next stage of the experiment involves the conversion of zero quantum coherence into transverse magnetization by a regular sequence of pulses of very small flip angle α, each pulse converting only a very small fraction of the available zero quantum coherence. It will be shown below that each α pulse destroys an amount of zero quantum coherence proportional to α^2, so that if α is too large there is a rapid decay of the detected signal, and the zero quantum lines appear broadened. In practice α was about 0.03 radians. There is then a delay τ_4 before the transverse magnetization is sampled. The transverse magnetization arising from zero quantum coherence appears initially as two antiphase components (section 1.2.4), and τ_4 allows relative precession of these vectors so that they become partly in phase. There is also a loss of signal during τ_4 due to defocussing effects in the inhomogeneous field and therefore only a rather short value of τ_4 is chosen, in practice 5 msec.

After a single sample point has been acquired, a 1-msec. gradient pulse is applied in order to destroy all transverse magnetization before the next α pulse. In order to prevent steady-state effects (35), these gradient pulses are amplitude-modulated by a pseudo-random sequence; in practice there was a constant component giving a line width of 3 kHz and an amplitude-modulated component with excursions of the same magnitude. The highest-frequency zero quantum coherence is of the order of the largest chemical shift difference in the coupled spin system, and this determines the lowest possible sampling frequency if aliasing is to be avoided. This is a further reason for employing shorter-than-optimum values of τ_4; it also means that weakly coupled spin systems with large values of δ/J will produce only weak zero quantum spectra by this method.

This process of pulsing and sampling is continued until 1024 data points are acquired, and this transient signal is then zero-filled to 4096 points. In addition to the oscillating signal derived from zero quantum coherence, this transient contains another component originating from diagonal elements of the density matrix, which grows monotonically by spin-lattice relaxation. A dc correction routine is used to avoid the step function discontinuity at the end of the transient signal. Gaussian weighting is then applied, and Fourier transformation gives the spectrum of zero quantum coherence.

To see the effect of a sequence of pulses, suppose that a spin system described by a density matrix σ experiences a regular sequence of pulses of very small flip angle α radians, applied along the x-axis of the rotating reference frame. The effect of a single pulse with flip angle α is then to create a new density matrix σ' which may be written as

$$\sigma' = \exp(-i\alpha\hat{F}_x)\sigma\exp(i\alpha\hat{F}_x) \qquad [4.9]$$

and, if expressed as a series expansion, may be approximated for small values of α by the first-order terms

$$\sigma' = (1 - i\alpha\hat{F}_x)\sigma(1 + i\alpha\hat{F}_x) \qquad [4.10a]$$

$$\sigma' = \sigma + i\alpha[\sigma,\hat{F}_x] + \alpha^2\hat{F}_x\sigma\hat{F}_x. \qquad [4.10b]$$

Provided that σ contains only density matrix elements that are diagonal or which represent zero quantum coherences, the term $[\sigma,\hat{F}_x]$ creates transverse magnetization without affecting the zero quantum coherences at all. (Off-diagonal non-zero quantum matrix elements are removed by the inhomogeneous field and the pulsed field gradients). The last term of Eq. [4.10b] and the higher-order terms in the series expansion of $\exp(i\alpha\hat{F}_x)$ in Eq. [4.9] cause the destruction of the zero quantum coherence by the α pulses. Consequently the amplitude of the transverse magnetization generated is proportional to α while the degree of destruction of the zero quantum coherence is proportional to α^2, provided that α is small. This effect is closely related to the flip-angle effect as described in section 2.3.3.

Zero quantum transitions in the vinyl group of triethoxyvinylsilane, which forms a convenient AMX spin system, were studied using the experimental scheme of Fig.4.12. Both the zero quantum spectrum and the conventional one-dimensional spectrum are shown in Fig.4.13. All six zero quantum transitions are clearly visible.

The author does not expect that this experiment will become of great practical importance; however, it shows an elegant demonstration of both the flip-angle effect and the behaviour of zero quantum coherence in a strongly inhomogenous magnetic field.

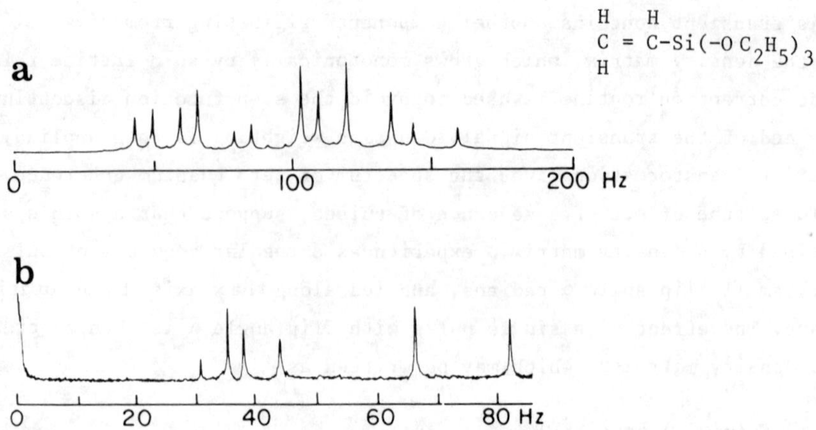

Fig.4.13 (a) The conventional high-resolution NMR spectrum of the vinyl protons in triethoxyvinylsilane. The three-spin system is sufficiently weakly coupled that three principal quartets can be picked out. (b) The spectrum of zero quantum frequencies (absolute value mode). Note the change of frequency scale. From ref.(34).

4.7 Discussion

Multiple quantum spectroscopy as described in this chapter is a clear demonstration of the potential power of two-dimensional spectroscopy. However, the practical usefulness of multiple quantum coherence in isotropic media still remains rather limited compared with the methods described in the previous chapters. Possible applications of multiple quantum coherence are:

1. Determination of connectivity,
2. Determination of relative signs of coupling constants,
3. Spectral simplification,
4. Separation of different spin systems,
5. Extra information for relaxation studies.

ad 1,2. Points 1 and 2 were already proposed in the early days of multiple quantum NMR. Since in those days double resonance experiments were used for the same purpose, and were simpler to understand by the average spectroscopist, multiple quantum NMR never found wide-spread popularity. Because both point 1 and 2 can be obtained with the correlation experiments discussed in chapter 2, they probably will not be a reason for a revival of interest in multiple quantum NMR.

ad 3. Spectral simplification can be achieved by studying a high order multiple quantum spectrum since the number of transitions decreases rapidly for increasing order higher than one. This property appears to be extremely useful in the study of dipolar coupling constants in molecules dissolved in liquid crystals (22,23). The simplification has a different character as the simplification obtained with the methods discussed in chapters 2 and 3.

ad 4. By selectively observing signals which originate from a certain order of transitions, using the special properties of multiple quantum transitions (section 4.3) to obtain this separation, it is possible to observe certain types of spin systems selectively in a conventional one-dimensional experiment. A practical application of this is used in the next chapter where ^{13}C - ^{13}C coupled spins are observed selectively, suppressing the much stronger signals due to isolated ^{13}C spins.

ad 5. As has been pointed out in several publications, multiple quantum relaxation can give useful extra information in motional studies (32,27,15). Especially in the study of correlations in fluctuations of the magnetic field at different sites in a molecule, relaxation rates of multiple quantum transitions are extremely helpful. Zero quantum relaxation rates are of particular interest because of the high reliability of the experimental results. Since motional studies using transverse relaxation phenomena in nuclear magnetic resonance are the work of only a limited number of scientists, this will be no reason for multiple quantum NMR to be introduced in the NMR cookbook of the chemist.

However, one has to realize that the development of pulsed multiple quantum NMR is rather recent and that new prospects may be found.

References

(1) W.A. Anderson, Phys. Rev. 104, 850 (1956).

(2) W.A. Anderson, R. Freeman and C.A. Reilly, J. Chem. Phys. 39, 1518 (1963).

(3) R. Freeman and W.A. Anderson, J. Chem. Phys. 39, 806 (1963).

(4) V.J. Kowalewski, Progress in NMR Spectroscopy 5, p.1, J.W. Emsley and J. Feeney and L.H. Sutcliffe (eds.), Pergamon Press, Oxford 1969.

(5) H. Hatanaka, T. Terao and T. Hashi, J. Phys. Soc. Japan 39, 835 (1975).

(6) A. Wokaun and R.R. Ernst, Chem. Phys. Lett. 52, 407 (1977).

(7) A. Wokaun, Ph.D. Thesis, Zürich 1978.

(8) S. Vega and A. Pines, J. Chem. Phys. 66, 5624 (1977).

(9) S. Vega, J. Chem. Phys. 68, 5518 (1978).

(10) A. Wokaun and R.R. Ernst, J. Chem. Phys. 67, 1752 (1977).

(11) R. Kaiser, J. Magn. Reson. 40, 439 (1980).

(12) L. Müller, J. Am. Chem. Soc. 101, 4481 (1979).

(13) A. Minoretti, W.P. Aue, M. Reinhold and R.R. Ernst, J. Magn. Reson. 40, 175 (1980).

(14) S. Emid, A. Bax, J. Konijnendijk, J. Smidt and A. Pines, Physica 96B, 333 (1979).

(15) G. Bodenhausen, Progr. in NMR Spect. 14, 137 (1981).

(16) R.R. Ernst and W.A. Anderson, Rev. Sci. Instr. 37, 93 (1966).

(17) G. Bodenhausen, R. Freeman and G.A. Morris, J. Magn. Reson. 23, 171 (1976).

(18) G.A. Morris and R. Freeman, J. Magn. Reson. 29, 433 (1978).

(19) W.S. Warren, S. Sinton, D.P. Weitekamp and A. Pines, Phys. Rev. Lett. 43, 1791 (1979).

(20) W.S. Warren, D.P. Weitekamp and A. Pines, J. Magn. Reson. 40, 581 (1980).

(21) A. Pines, S. Vega, D.J. Ruben, T.W. Shattuck and D.E. Wemmer, Proc. IV Ampere Int. Summer School, Pula, Yugoslawia 1976.

(22) A. Pines, D.E. Wemmer, J. Tang and S. Sinton, Bull. Am. Phys. Soc. 23, 21 (1978).

(23) T.W. Shattuck, Ph.D. Thesis, Univ. Calif. Berkeley 1978.

(24) M. Tinkham, Group Theory and Quantum Mechanics, chapter 2, Mc Graw-Hill, New York 1964.

(25) A.A. Maudsley, A. Wokaun and R.R. Ernst, Chem. Phys. Lett. 55, 9 (1978).

(26) A. Bax, P.G. de Jong, A.F. Mehlkopf and J. Smidt, Chem.Phys. Lett. 69, 567 (1980).

(27) G. Bodenhausen, R.L. Vold and R.R. Vold, J. Magn. Reson. 37, 93 (1980).

(28) A. Bax, R. Freeman, T.A. Frenkiel and M.H. Levitt, J. Magn. Reson. 43, 478 (1981).

(29) T.A. Frenkiel, M.H. Levitt and R. Freeman, J. Magn. Reson. 44, 409 (1981)

(30) G. Bodenhausen, J. Magn. Reson. 34, 357 (1979).

(31) R.L. Vold, R.R. Vold, R. Poupko and G. Bodenhausen, J. Magn. Reson. 38, 141 (1980).

(32) A. Wokaun and R.R. Ernst, Mol. Phys. 36, 317 (1978).

(33) H.Y. Carr and E.M. Purcell, Phys. Rev. 94, 630 (1954).

(34) A. Bax, A.F. Mehlkopf, J. Smidt and R. Freeman, J. Magn. Reson. 41, 502 (1980).

(35) R. Freeman and H.D.W. Hill, J. Chem. Phys. 54, 301 (1971).

5 CARBON - CARBON COUPLING OBSERVED IN NATURAL ABUNDANCE SAMPLES

5.1 Introduction

In this chapter a number of experiments is discussed which all have in common that signals which originate from homonuclear coupled carbon-13 spins are detected selectively, suppressing the much stronger signals from isolated carbon - 13 nuclei. The experiments are all rather recent and not much detailed literature is available yet. Therefore a separate chapter is used for the description of these experiments, although all principles on which the new methods rely have been discussed in the previous chapters.

Carbon-carbon scalar coupling constants contain a wealth of information about the structure and conformation of carbon containing molecules. The information they provide is complementary to that obtained from proton-proton coupling constants, and is often unique because carbon coupling constants are directly influenced by the bonding and substituent effects in the carbon skeleton of organic molecules. However, there appears to be a difficult practical problem in measuring these coupling constants. This difficulty arises because the ^{13}C isotope is rare; it has an abundance of only approximately 1%. Hence, the abundance of ^{13}C - ^{13}C pairs is only 0.01% and the intensity of the corresponding resonances will be a factor 200 lower than the intensity of signals from the isolated ^{13}C spins. Since homonuclear direct ^{13}C coupling constants are rather small, 30-45 Hz for saturated carbons, these weak satellites are hard to detect in the flanks of the 200 times stronger isolated ^{13}C signal. The problem is not only the sensitivity or the dynamic range of the signal, but also the presence of a jumble of other weak lines from spinning side bands, from incomplete proton decoupling or simply from small amounts of impurities. As an example in Fig.5.1, one of the ^{13}C resonances of piperi-

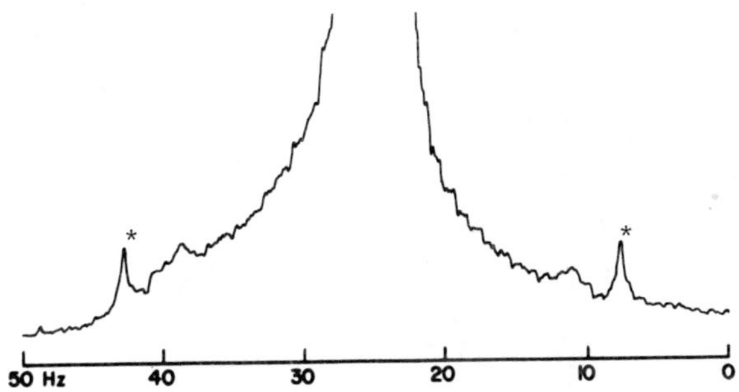

Fig.5.1 Carbon-13 resonance line of carbon C_4 in piperidine in a conventional spectrum recorded at 50 MHz. The satellites due to homonuclear coupling with C_3 are marked with an asterisk.

dine is shown with the two satellite signals due to homonuclear ^{13}C coupling in its flanks.

Especially at high magnetic fields it becomes very difficult to decouple the protons properly and to obtain a narrow base for the relatively strong resonance due to isolated ^{13}C nuclei. Therefore the solution forced upon the chemist has been to synthesize compounds labelled with ^{13}C at the site of interest, and sometimes even labelled at two sites. This procedure is difficult, time-consuming and expensive, and restricts the measurements that can be made to those sites which are coupled to the labelled atoms. In spite of this, a large amount of information has been built up on the values of carbon-carbon coupling constants (1) using this isotope substitution. Experiments performed by L. Ernst and Wray (2), using no isotope substitution, but detecting direct and long range satellites in conventional spectra of natural abundance samples of monosubstituted benzene compounds, should be considered as experimentally extremely difficult, requiring excellent homogeneity, high decoupling power, and properly chosen compounds.

The method of measuring carbon-carbon coupling constants described in this thesis is completely different from both methods mentioned above; the new method relies on suppressing the main signals of isolated ^{13}C nuclei, retaining the interesting satellite lines. As will be explained in the next paragraph, the suppression is achieved by using phase properties of double quantum coherence (section 4.3). The experiments are essentially one-dimensional; the idea, however, originates from the multiple quantum spectroscopy as presented in chapter 4. This one-dimensional method (3) will further be referred to as Inadequate (Incredible Natural Abundance Double Quantum Transfer Experiment). In paragraphs 5.4 and 5.5 it will be shown that the two-dimensional variations of the Inadequate experiment can be extremely useful for the detection of long range couplings and for assignment purposes.

All experiments described in this chapter operate by cancelling uncoupled resonances to reveal the satellites; it should be noted that the sensitivity of the coupled resonances is not enhanced, making it necessary to use strong solutions and long accumulation times.

During all experiments discussed in this chapter broad-band high power proton decoupling is used throughout.

All spectra shown in this chapter are recorded on a Varian XL 200 spectrometer, operating at 50 MHz.

5.2 The Inadequate experiment

The dilution of carbon-13 spins in nature is not without advantage: the natural abundance carbon-carbon coupled spectrum consists of a number of subspectra from isotopomers of the same molecule, each containing two ^{13}C atoms, with the result that all these subspectra are either of the AX or the AB type. The intensity of AMX type spectra, due to molecules with three ^{13}C atoms will be a factor 200 weaker, and can therefore be neglected. This spectral simplicity facilitates the analysis and assignment of resonances in the coupled spectrum. The problem remaining is the suppression of the strong signals from the isolated ^{13}C spins. The key to the separation method is that only in the case of coupled ^{13}C spins it is possible to generate double quantum coherence. The special phase properties as discussed in sections 4.3 and 4.4 can be used to select signals which originate from double quantum coherence only, and hence belong automatically to coupled ^{13}C spins (3).

For the excitation of double quantum coherence the offset-independent excitation (section 4.2, Fig.4.3g) $(\pi/2)_x - \tau - (\pi)_y - \tau - (\pi/2)_x$ is used. Neglecting pulse imperfections and relaxation during the time intervals τ, only two components are created: longitudinal magnetization along the -z-axis and double quantum coherence.

Using a density matrix calculation (section 1.2.3, App.I) it can be found that the amount of double quantum coherence created is for a weakly coupled spin system proportional to $\sin(2\pi J_{CC}\tau)$ and hence the condition for optimum transfer into double quantum coherence is given by

$$\tau = (2n + 1)/4J_{CC} \qquad n = 0,1,2,3,... \qquad [5.1]$$

Due to pulse imperfections and relaxation an amount of single quantum coherence is also created, which can be considerably larger than the amount of double quantum coherence from the isolated $^{13}C - {}^{13}C$ pairs.

Another $\pi/2$ pulse with phase ϕ is applied immediately afterwards, and converts all three components into transverse magnetization. The complete pulse sequence is set out in Fig.5.2, where Δ is a very short delay, (in practice 2 μsec.), needed for resetting the phase of the radiofrequency pulse. By performing several experiments with different phase ϕ for the detection pulse, signals can be selected which originate only from double quantum coherence, as described in section 4.4.3. In Table 5.1 the phase of the three components originating from longitudinal magnetization and from single and double quantum coherence is set out for four phases ϕ of the mixing pulse.

The phase Ψ of the receiver is made to follow the signal originating from

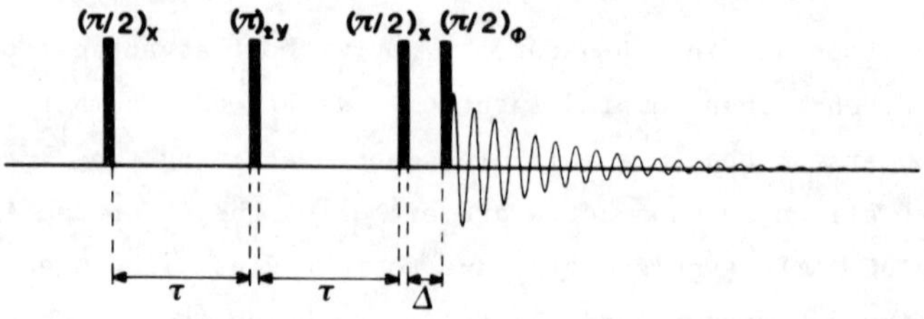

Fig.5.2 Pulse sequence of the Inadequate experiment. The phases ϕ and Ψ are cycled according to Table 5.1.

Table 5.1 Phase of the detection pulse ϕ and the receiver reference phase Ψ, compared with the phase of the three components originating from longitudinal magnetization (LM), single quantum (SQ), and double quantum (DQ) coherence.

ϕ	LM	SQ[1]	DQ[1]	Ψ
+x	-y	+x	+x	+x
+y	+x	+y	-y	-y
-x	+y	+x	-x	-x
-y	-x	+y	+y	+y

[1]Phase of only one of the two doublet components (see e.g. App.I).

double quantum coherence, causing cancellation for the other components as soon as four experiments have been performed. This four-step experiment can give quite good suppression of the uncoupled ^{13}C signals, but some spurious signal at the frequency of the strong resonance will be left. This spurious signal is due to spectrometer instabilities, pulse imperfections (H_1-inhomogeneity, off-resonance effects), to imperfections in the phase shift of the $\pi/2$ and the π pulses, and to errors in the phase shifts in receiver channels and inequality of the audiofrequency filters in the quadrature detection system. In order to eliminate the spurious signals due to these imperfections, the following extensions of the phase-cycling proposed in Table 5.1 can be used:

1. The entire four-step experiment is repeated four times with phases of all pulses and the receiver reference phase incremented together by $\pi/2$ each time, giving rise to a 16-step experiment. This extra cycling is analogous to the Cyclops modification in conventional Fourier transform NMR experi-

ments (4).

2. The phase of the first pulse can be inverted if the receiver reference phase is inverted as well. This extends the experiment to a 32-step sequence.

3. The phase of the π pulse can be cycled through all four phases (x,y,-x,-y), provided that the receiver reference phase is inverted each time the phase of the π pulse is incremented. This finally gives rise to a 128-step cycle, which is the maximum number of steps which can be obtained with the four-pulse experiment, and hence gives optimum suppression.

The order of points 1,2 and 3 is the order of importance for good suppression as found empirically on a Varian XL 200 spectrometer. If the step number N is given by

$$N = n_1 + 4 \times n_2 + 16 \times n_3 + 32 \times n_4 \qquad 0 \leqslant n_1, n_2, n_4 \leqslant 3 \text{ and } 0 \leqslant n_3 \leqslant 1,$$

where n_1, n_2, n_3 and n_4 are integer numbers, the phase of the corresponding pulses and of the receiver is given below. Only one combination of n_1, n_2, n_3 and n_4 is possible for each value of N and the phases of the first $\pi/2$ pulse ϕ_1, of the π pulse ϕ_2, of the second $\pi/2$ pulse ϕ_3, of the final $\pi/2$ readpulse ϕ, and the receiver reference phase Ψ, are given by the following expressions for a certain step in the 128-step sequence:

$$\begin{aligned} \phi_1 &= (n_2 + 2 \times n_3) \times \pi/2 \\ \phi_2 &= (n_4 + n_2) \times \pi/2 \\ \phi_3 &= (n_2) \times \pi/2 \\ \phi &= (n_1 + n_2) \times \pi/2 \\ \Psi &= (1 - n_1 + n_2 + 2 \times n_3 + 2 \times n_4) \times \pi/2. \end{aligned} \qquad [5.2]$$

A phase $\phi_n = 0$ corresponds to a pulse along the x-axis, a phase $\phi_n = \pi/2$ corresponds to a pulse along the y-axis, etc.

As an example the technique was applied to the natural abundance ^{13}C spectrum of piperidine. There are four observable carbon-carbon couplings, $^1J_{2,3} = 35.2$, $^3J_{2,3'} = 1.7$, $^2J_{2,4} = 2.6$ and $^1J_{3,4} = 33.0$ Hz. The first step is to set the condition for optimum transfer into double quantum coherence for the two direct couplings according to Eq. [5.1], knowing that the direct coupling of saturated carbons is of the order of 35 Hz. This allows $^1J_{2,3}$ and $^1J_{3,4}$ to be evaluated, and with the assumption that the long range couplings would be of the order of 5 Hz, both the direct and the long range couplings may be observed in the same spectrum by setting $\tau = 51$ msec., and letting n from Eq. [5.1] be equal to n = 3 and n = 0 respectively. The conditions for satisfying Eq. [5.1] are not critical when n = 0, and the signals corresponding to long range couplings will not be more than 30% below maximum intensity if the actual long range coupling differs less than 50% from the 5 Hz chosen. Fig. 5.3 shows 50 Hz sections from the spectrum of piperidine centered on the three carbon shifts which are equal 47.9, 27.8 and 25.9 ppm for C_2, C_3 and C_4 respectively. All three sections are taken from a single experiment, taking an accumulation time of approximately five hours. Both $^1J_{2,3}$ and $^1J_{3,4}$ are obtained from AB type spectra with significant displacements of the centres of the doublets from the chemical shift frequencies, and with noticable asymmetry.

As the net magnetization originating from double quantum coherence has to be zero after the non-selective detection pulse, the doublet components have to be in antiphase immediately after this detection pulse (section 1.2.4). Whether the low field doublet component has positive or negative intensity (up-down or down-up) depends upon whether n from Eq. [5.1] is even or odd. In the piperidine spectra the doublets due to long range couplings (n = 0) are all up-down while those from direct couplings (n = 3) are all down-up.

When there has been optimal transfer into double quantum, the spectra from ^{13}C - ^{13}C coupled spins appear with the same sensitivity as in the conven-

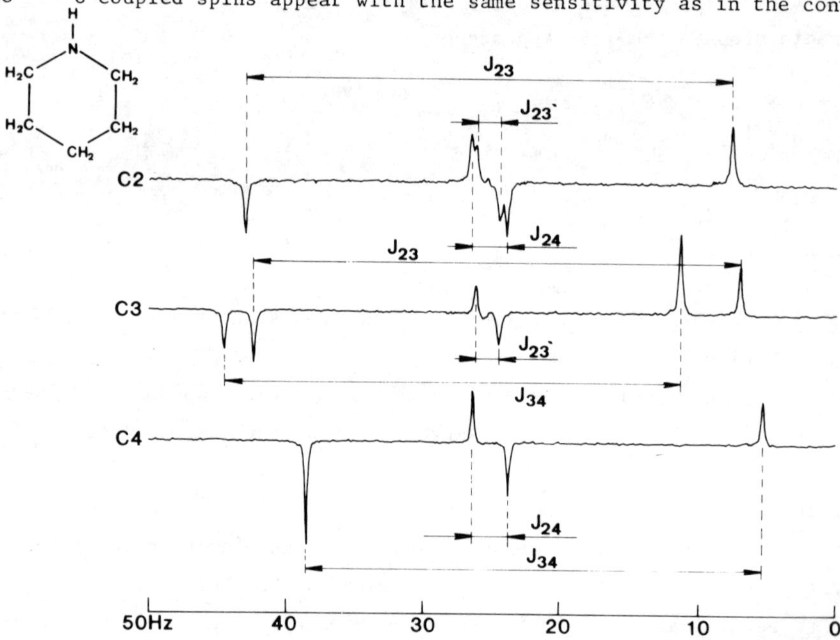

Fig.5.3 Sections from the 50 MHz carbon-13 spectrum of piperidine, showing direct and long range couplings. Note the effective suppression of the signals from the much more abundant molecules with isolated ^{13}C nuclei. From ref.(3).

tional spectrum. Signals from impurities and modulation artefacts are effectively suppressed, leaving a clean spectrum of the desired isotopomers. The suppression of the non-coupled ^{13}C in Fig.5.3 is over a factor 1000.

As found by Frenkiel (5) the sensitivity of the method decreases rapidly if the off-resonance effect becomes noticeably large ($\gamma H_1 < 5\Delta\omega$, where $\Delta\omega/2\pi$ is the offset frequency of a certain resonance). In order to overcome this problem for wide spectra, where e.g. methyl groups are coupled to a carbonyl, the use of composite pulses (6) for all pulses in the sequence is recommended.

The delay time after every four-pulse step in the experiment can be optimized for sensitivity by setting the sum of acquisition times and this delay time equal to $1.3T_1$, where T_1 is the longitudinal relaxation time of the ^{13}C spin (7). In the case where the longitudinal relaxation times of two coupled spins differ by a large amount, e.g. a protonated carbon coupled to a quaternary carbon, this time can be set to 1.3 times the shorter T_1, giving satellites for both carbon sites with about half the intensity.

5.3 Optimization in the case of strong coupling

In practice quite a few of the ^{13}C - ^{13}C spectra are strongly coupled. Optimization of sensitivity using Eq. [5.1] is then no longer applicable (8). It is the purpose of this section to calculate how to set the pulse timing for good sensitivity when there are strongly coupled pairs of ^{13}C spins. A density matrix calculation, using the equations given in section 1.2.3, is now necessary to calculate how much double quantum coherence is created after the $(\pi/2)_x$ - $(\pi)_y$ - $(\pi/2)_x$ pulse sequence. For the density matrix σ' after this pulse sequence one finds

$$\sigma' = \hat{R}_x(-\pi/2)\exp(-i\hat{H}\tau)\hat{R}_y(-\pi)\exp(-i\hat{H}\tau)\hat{R}_x(-\pi/2) \times \times \sigma_{eq}\hat{R}_x(\pi/2)\exp(i\hat{H}\tau)\hat{R}_y(\pi)\exp(i\hat{H}\tau)\hat{R}_x(\pi/2), \quad [5.3]$$

where the rotation operator is given in Eq. [1.15] . A matrix representation for this operator in the case of an AB spin system is given in ref.(9). A further pulse along an axis in the xy-plane which makes an angle ϕ with the positive x-axis converts the double quantum coherence, present in σ', into detectable single quantum coherence described by the density matrix σ'':

$$\sigma'' = R_\phi(-\pi/2)\sigma' R_\phi(\pi/2). \quad [5.4]$$

Using Eq. [1.26] the detected magnetization is now given by

$$M_{tr}(t_1,t_2) = C\ \mathrm{Tr}(\hat{F}_x + i\hat{F}_y)\sigma(t), \quad [5.5a]$$

where

$$\sigma(t) = \exp(-i\hat{H}t)\sigma''\exp(i\hat{H}t). \quad [5.5b]$$

From explicit calculation, using Eqs. [5.3] - [5.5], it follows that if the four transition frequencies observed in the conventional AB spectrum are written ω_{12}, ω_{34}, ω_{13} and ω_{24} in order of increasing frequency, and if $\tan(2\theta) = J/\delta$ then the sum I of the absolute intensities of the four observed transitions is given by

$$I = s(1 + s)\sin\{(\omega_{13} - \omega_{34})\tau\} - 2c^2\sin\{(\omega_{34} - \omega_{12})\tau\} + \\ + s(1 - s)\sin\{(\omega_{24} - \omega_{12})\tau\}, \quad [5.6]$$

where $s = \sin(2\theta)$ and $c = \cos(2\theta)$. The second term in Eq. [5.6] is the term present in the weakly coupled limit.

Eq. [5.6] can be easily verified experimentally by making τ a variable ($=\frac{1}{2}t_1$) as shown in Fig.5.4, and transforming the resulting data matrix $s(t_1,t_2)$ to give a two-dimensional spectrum $S(\omega_1,\omega_2)$.

A schematical diagram of such a spectrum is shown in Fig.5.5. A cross-section through this spectrum parallel to the F_1-axis as indicated in Fig.5.5 gives a one-dimensional spectrum consisting of six lines, three with positive frequencies and three with equal negative frequencies, having the relative intensities indicated by Eq. [5.6].

The methyl and quaternary carbons in tetramethyladamantane provide a suitable example for testing Eq. [5.6], since the relative chemical shift is 92.3 Hz and the direct coupling is 37.0 Hz, giving $2\theta = 0.38$ radians. The experimental spectrum, obtained from a cross-section through the 2D spectrum, is illustrated in Fig.5.6 and is compared with the predicted stick spectrum, calculated from Eq. [5.6], showing good agreement. The observed frequencies

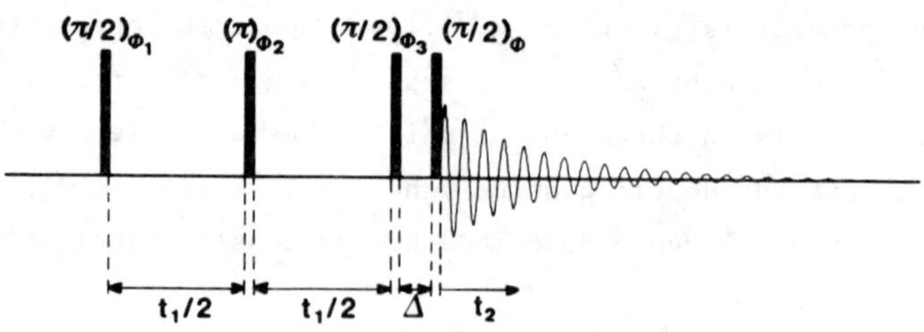

Fig.5.4 Two-dimensional version of the Inadequate experiment. The phases ϕ_1, ϕ_2, ϕ_3, ϕ and Ψ are cycled according to Eq. [5.2].

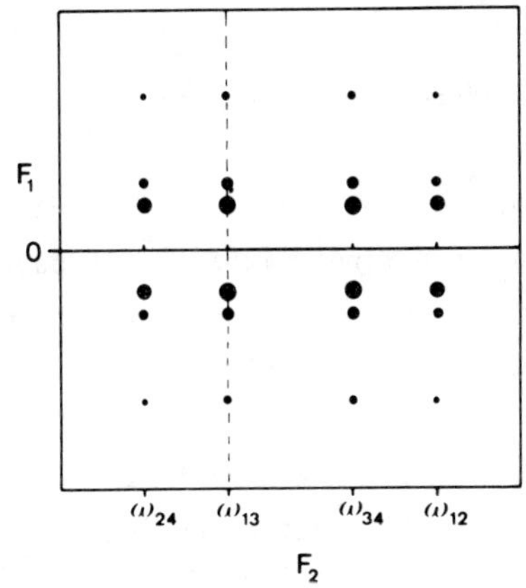

Fig.5.5 Schematic 2D spectrum of a pair of strongly coupled ^{13}C spins using the pulse sequence of Fig.5.4. In Fig.5.6 a cross-section at the position indicated by the dashed line is shown.

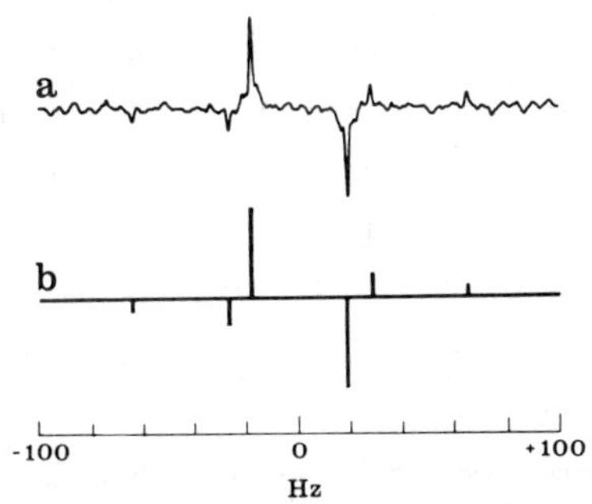

Fig.5.6 (a) A cross-section through the 2D ^{13}C spectrum of the AB spin system in tetramethyladamantane . (b) The predicted spectrum calculated from Eq. [5.6] using the measured values of J and δ. From ref.(8).

and relative intensities are set out in Table 5.2. Expressions identical to

Table 5.2 Calculated and observed relative intensities in the F_1 spectrum of the AB system in tetramethyladamantane.

Line frequency	Calculated intensity	Observed intensity
$(\omega_{34} - \omega_{12})/2\pi$; ±18.5 Hz	1.00	1.00
$(\omega_{13} - \omega_{34})/2\pi$; ±27.5 Hz	0.29	0.27
$(\omega_{24} - \omega_{12})/2\pi$; ±64.6 Hz	0.14	0.16

Eq. [5.6] have been calculated for the frequencies and relative intensities in a homonuclear J-spectrum (section 3.3) of an AB spin system (10). The only difference is a change in sign in the first term in Eq. [5.6] attributable to the phase inversion which occurs in the present experiment when a line appears at a negative frequency.

The density matrix calculation using Eqs. [5.3] - [5.5] predicts that the relative intensities of the four resonance lines within an AB spectrum remain exactly the same when detected via double quantum coherence. The overall intensity, however, is a function of $J_{CC}\tau$ and the strength of the coupling δ/J_{CC}.

Since Eq. [5.6] contains three terms oscillating at different frequencies, it is not possible to give a simple prescription for setting τ for optimal signal intensity.

The third term (oscillating at the frequency separation of the outer lines of the AB quartet) may nearly always be neglected in comparison with the other terms; consequently it is the sum of these first two terms which effectively determines the optimum setting of τ. At the condition $\delta/J = \sqrt{3}$ the inner lines of the AB pattern are separated by exactly J and both terms are oscillating with the same frequency. For more strongly coupled cases, the second term oscillates at a higher frequency and there is a local maximum of signal intensity close to the condition of Eq. [5.1]. It is therefore possible to relate the optimum setting of τ to J rather than the separation of the inner lines of the AB quartet.

The intensity near this local maximum may not be very high, and for cases where the coupling is stronger than the condition $\delta/J = 2.8$ there can be a significant advantage in setting $J\tau = 3/4$ rather than the simpler condition $J\tau = 1/4$. This is most easily appreciated by a graphical presentation based on numerical evaluation of Eq. [5.6] as illustrated in Fig.5.7. Several gene-

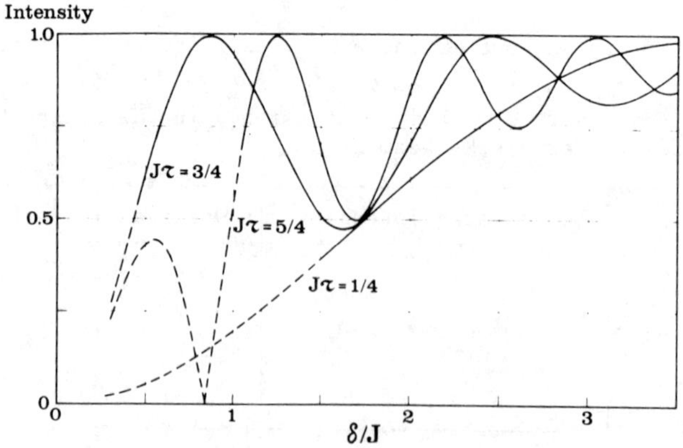

Fig. 5.7 Calculated relative intensities for the ^{13}C - ^{13}C satellite spectra as a function of the strength of the coupling in an AB spin system. Intensity 1 corresponds to the intensity of the satellites in a conventional spectrum after a $\pi/2$ pulse. From ref.(8).

ral conclusions can be drawn from this graph. For all weakly coupled systems $J\tau = 1/4$ is the best choice, and it remains reasonably good down to the condition $\delta/J = \sqrt{3}$. For strong coupling better signal intensity can be achieved with the condition $J\tau = 3/4$, although this is three times as sensitive to mismatch. In some limited regions of the graph the condition $J\tau = 5/4$ (or even higher multiples) can give stronger signals. The curves for the three conditions $J\tau = 1/4, 3/4, 5/4$ are plotted in Fig.5.7, the broken lines indicating that in these regions there may not be a suitable local maximum of signal intensity given approximately by Eq. [5.1]. Such settings of τ should therefore be avoided. Note that $J\tau = 1/4$ is an unsatisfactory choice for strongly coupled spins. For extremely strong coupling, where δ/J is less than about 0.5, a different strategy should be adopted since the first term of Eq. [5.6] dominates completely. The timing should then be adjusted so that $(\omega_{13} - \omega_{34})\tau = \pi/2$, which requires a knowledge of the separation of the inner lines of the AB quartet, that is, an estimate of the chemical shift difference as well as J.

5.4 A versatile two-dimensional method for the investigation of long range couplings

The principle restriction on the generality of the Inadequate experiment arises from the condition for optimum transfer into double quantum coherence (Eq. [5.1]) as described in section 5.2. Optimum transfer requires either an estimate of J_{CC} or the trial and error adjustment of the τ delay. If the main interest is in direct couplings, an estimate of J_{CC} and a corresponding choice of τ is rather straightforward. However, in this case of long range couplings it is much harder to predict which value of J_{CC} to expect, since values vary between 0 and 15 Hz. For the routine application to problems where no a priori estimates of the couplings can be made a more general method would clearly be useful. In this paragraph a variation on the Inadequate experiment will be discussed, which utilizes the concept of two-dimensional spectroscopy to allow τ to become a variable parameter in this experiment (Fig.5.4) (11). In principle this should exhibit a sensitivity comparable with that of the Inadequate experiment with optimized τ delays. The period 2τ becomes the well-known evolution period t_1. For each value of t_1 an Inadequate experiment gives a satellite spectrum, with the intensities of the satellites proportional to

$$I(t_1) = \sin(\pi J_{CC} t_1)\exp(-t_1/T), \qquad [5.7]$$

provided that the coupling is weak, as is usually the case with long range couplings. Because of the refocussing effect of the π pulse, the decay time constant T of this signal can approach the spin-spin relaxation time T_2 under favourable conditions. A Fourier transformation of this amplitude modulation with respect to t_1 will give resonance lines at $\pm J_{CC}/2$ in the F_1 dimension. Because of the refocussing effect of the π pulse, the resolution in the F_1 dimension may be considerably enhanced in comparison with conventional experiments. Thus, although the $^{13}C - ^{13}C$ splittings appear in the two-dimensional spectrum in both frequency dimensions, there is a marked advantage in observing them in the F_1 dimension. This is conveniently achieved by selecting the appropriate traces from the complete two-dimensional spectrum, the equivalent of taking cross-sections through the two-dimensional spectrum at fixed values of F_2. In general it is not necessary to plot the two-dimensional spectrum. Since for each carbon site there are two satellites with opposite intensities (section 5.2), both modulated with respect to t_1 by the same frequency, sensitivity can be enhanced by a factor $\sqrt{2}$, by calculating the difference of the two corresponding cross-sections. Since the signal shows pure amplitude modulation, sensitivity can be further enhanced by performing a real Fourier transformation with respect to t_1 or by replacing the dispersive part of the individual t_1-satellite spectra by zeroes before performing complex Fourier transformation with respect to t_1 (section 1.8). This latter approach gives the familiar up-down pattern for the satellites in the F_1 cross-sections.

In order to illustrate the technique, an investigation was made of the long range couplings in tetramethyladamantane, chosen because the alternative approach, specific isotopic enrichment, is particularly difficult in ring compounds. As can be appreciated from Fig.5.8 symmetry considerations limit the observable long range C - C couplings to four:

$^2J(CH_3\text{-}C\text{-}CH_2)$, $^3J(CH_3\text{-}C\text{-}CH_2\text{-}C)$, $^3J(CH_2\text{-}C\text{-}CH_2\text{-}C)$, $^4J(CH_3\text{-}C\text{-}CH_2\text{-}C\text{-}CH_2)$.

Three of these couplings turn out to be less than 1 Hz, making this a stringent test of the method.

The number of increments in the t_1 dimension is a compromise between the requirement for adequate digitization in the F_1 dimension and the desire to minimize the overall length of the experiment; in practice 32 increments were used. The overall data matrix consisted of 32 × 4096 words. The signals (t_2) were weighted with a Gaussian function to give a line width in the F_2 dimension of 0.35 Hz. Thus although in principle each splitting should appear on a different horizontal (F_1) trace, in practice as many as three splittings appeared together, because of overlap of the signals in the F_2 dimension.

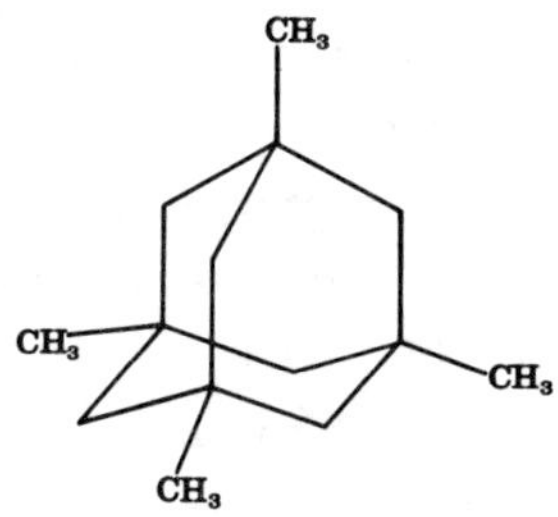

Fig.5.8 Tetramethyladamantane.

The same splitting information occurs on both sides of the ^{13}C chemical shift frequency, the spectra having antiphase intensities. This can be seen in Fig. 5.9 which shows a small section taken from the two-dimensional spectrum encompassing the resonances from the methyl carbon sites. Trace (a) shows an antiphase doublet due to $^{3}J(C\text{-}CH_2\text{-}C\text{-}CH_3)$ while trace (d) carries a similar doublet with the intensities inverted. Consequently traces (a) and (d) were combined in antiphase to give a resultant spectrum with improved signal-to-noise. Similarly trace (b) was inverted and then combined with trace (c) to give a spectrum illustrating the smaller splittings $^{2}J(CH_3\text{-}C\text{-}CH_2)$ and

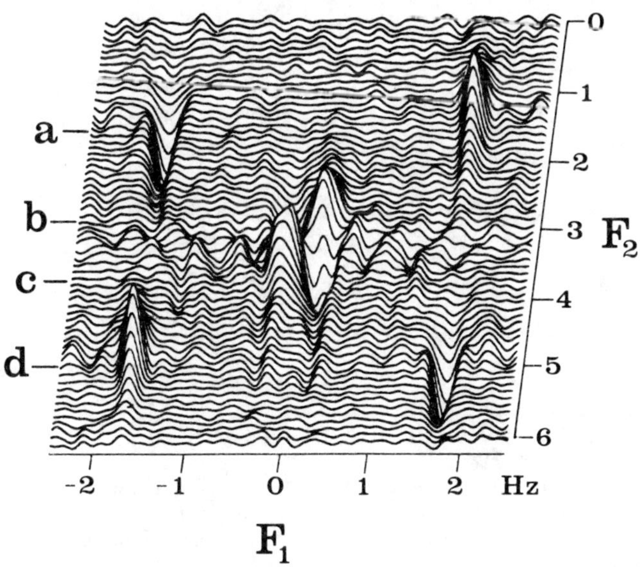

Fig.5.9 A small section of the two-dimensional ^{13}C spectrum of tetramethyladamantane, approximately centred on the methyl carbon chemical shift frequency. Traces (a) and (d) carry splittings due to $^{3}J(CH_3\text{-}C\text{-}CH_2\text{-}C)$, while traces (b) and (c) carry splittings due to both $^{2}J(CH_3\text{-}C\text{-}CH_2)$ and $^{4}J(CH_3\text{-}C\text{-}CH_2\text{-}C\text{-}CH_2)$ which are not resolved in this diagram. From ref.(11)

$^4J(CH_3\text{-}C\text{-}CH_2\text{-}C\text{-}CH_2)$ which are not quite resolved in this spectrum.

In this way five critical spectra were extracted from the two-dimensional data; they are set out in Fig.5.10. The sensitivity achieved by an overnight run was sufficient to permit the application of a resolution enhancement function in the t_1 dimension, reducing the line width in F_1 from 0.2 to 0.1 Hz. This allows all four long range splittings to be resolved. Each splitting appears twice in these spectra, which helps to confirm their assignment, the only ambiguity being between $^2J(CH_2\text{-}C\text{-}CH_3)$ and $^4J(CH_2\text{-}C\text{-}CH_2\text{-}C\text{-}CH_3)$; these were tentatively assigned on the assumption that $|^2J|>|^4J|$. The results are set out in Table 5.3, along with the two direct couplings observed in a separate experiment.

By removing the necessity of estimating τ values, the proposed extension of the original ^{13}C - ^{13}C coupling experiment makes the application to un-

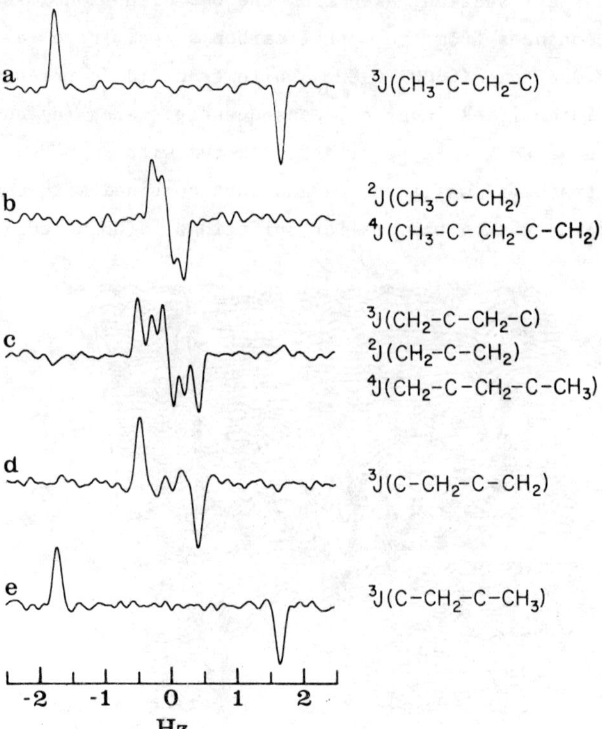

Fig.5.10 Spectra showing the long range ^{13}C - ^{13}C couplings in tetramethyladamantane obtained from sections through the two-dimensional spectrum parallel to the F_1-axis. A resolution enhancement function has been used. Traces (a) and (b) originate from the methyl groups, (c) from the methylene groups, and (d) and (e) from the quaternary sites. A small artefact near the centre of trace (d) is attributable to the failure to allow complete spin-lattice relaxation between sections of the experiment. From ref.(11).

Table 5.3 Carbon-carbon couplings in tetramethyladamantane (Hz)

$^{1}J(C\text{-}CH_3)$	$= 37.00 \pm 0.05$
$^{1}J(C\text{-}CH_2)$	$= 32.23 \pm 0.05$
$^{2}J(CH_3\text{-}C\text{-}CH_2)$	$= 0.45 \pm 0.03$[a]
$^{3}J(CH_3\text{-}C\text{-}CH_2\text{-}C)$	$= 3.38 \pm 0.03$
$^{3}J(CH_2\text{-}C\text{-}CH_2\text{-}C)$	$= 0.89 \pm 0.03$
$^{4}J(CH_3\text{-}C\text{-}CH_2\text{-}C\text{-}CH_2)$	$= 0.14 \pm 0.03$[a]

[a] Assigned on the assumption that $|^{2}J|>|^{4}J|$.

kwown long range couplings quite general. The attainable sensitivity is expected to be comparable with the one-dimensional experiment, particularly since it is possible to combine pairs of traces carrying the same information. The resolution may be significantly improved through the refocussing effect of the 180° pulse at the midpoint of the evolution period.

5.5 ^{13}C - ^{13}C double quantum frequencies used for assignment

Assignment in ^{13}C spectra can sometimes be very difficult and require a knowledge of isotope substitution effects or of the interpretation of longitudinal relaxation times. Quite a few false assignments have been made in the past. In this paragraph an assignment method is presented which uses the double quantum frequency of directly coupled ^{13}C - ^{13}C pairs for assignment purposes, identifying adjacent carbons in an unambiguous manner.

Each carbon may be directly coupled to as many as four other sites, and since the coupling constants are often very similar in magnitude, assignment to specific pairs of resonances cannot always be made on the basis of coupling constants alone. However, the couplings can be identified by means of a different criterion - the frequency of the double quantum coherence, which is equal to the sum of the chemical shifts of the two carbon sites, measured with respect to the transmitter frequency (section 4.2.3). Each pair of carbon spins has in most cases a unique double quantum frequency, since the double quantum frequencies are spread over a spectral region, almost twice as wide as the conventional ^{13}C spectral width. So the problem remaining is the detection of the ^{13}C - ^{13}C double quantum frequencies. This can be easily done in a two-dimensional experiment, using the pulse sequence of the Inadequate experiment, but letting the delay Δ now become the evolution period t_1 (Fig.5.11a). The delay τ is optimized for the detection of direct couplings (τ = 6 msec.). The satellite spectra obtained for each value of t_1 will now be modulated in amplitude with respect to t_1 by the double quantum frequencies. An explicit density matrix calculation, showing this, is given in Appendix I. A Fourier transformation with respect to t_1 will then produce a two-dimensional spectrum with along the F_1-axis the double quantum frequencies, and along the F_2-axis the satellite spectra as obtained with the Inadequate experiment. This method has been used to establish the connectivity of the carbon atoms of 5α - androstane (12). Since the signals are purely amplitude-modulated, sensitivity can be improved by performing a real Fourier transformation with respect to t_1, of the absorption part of the spectra obtained after the first Fourier transformation with respect to t_2 (section 1.5.2). A disadvantage of the amplitude modulation is that the sign of the double quantum frequency cannot be determined. This problem can readily be circumvented by placing the transmitter frequency at either the high-field or low-field side of the spectrum, ensuring that all double quantum frequencies have the same sign, but this increases the size of the data matrix required by a factor four. Since lack of digitization in the two-dimensional spectrum will be a problem anyway, this is not permissible.

As explained in section 4.4.3, the amplitude modulation by the double quantum frequency can be converted into phase modulation if either the coherence transfer echo or the antiecho is detected selectively. This can be done by combining the results of the experiment of Fig.5.11a with another similar experiment which has an extra $(\pi/4)_z$ pulse in the evolution period (Fig.5.11b). The composite $(\pi/4)_z$ pulse consists of a $(\pi/2)_{-x} - (\pi/4)_{-y} - (\pi/2)_x$ (13,14), and increases the phase of the double quantum coherence by $\pi/2$ (section 4.4.3). As the scheme of Fig.5.11a detects the imaginary component of the double quantum coherence (App.I), the scheme of Fig.5.11b detects the real component. If the receiver phase in the sequence of Fig.5.11b is decremented by 90° compared with sequence 5.11a, adding the two amplitude-modulated signals generates a phase-modulated signal, representing the coherence transfer echo and allowing the sign of the double quantum frequency to be determined (section 1.5.1). This is analogous to the tricks used in homonuclear (15) and heteronuclear (16) shift correlation and described in sections 2.2.3 and 2.3.2. It is convenient to insert the composite three pulse $(\pi/4)_z$ sequence immediately after the second $\pi/2$ pulse in the experiment, which creates the double quantum coherence. This allows four pulses to be consolidated into two:

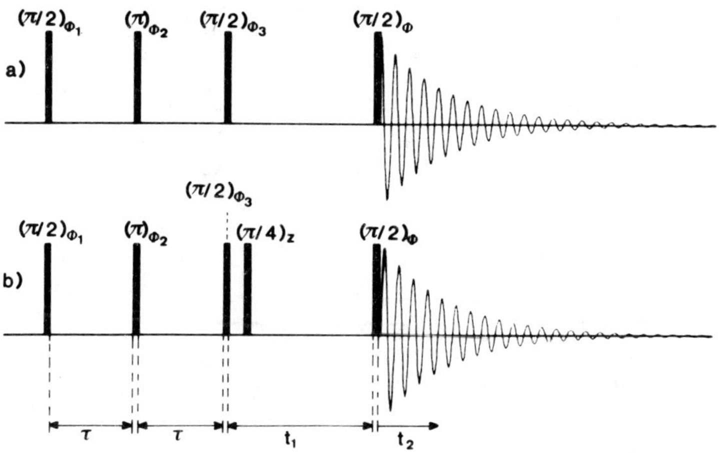

Fig.5.11 Two stages of the sequence used to detect carbon-13 satellites in carbon-13 spectra while relating them to the characteristic double quantum frequencies. The phase angles ϕ_1, ϕ_2, ϕ_3, ϕ and the receiver phase Ψ are cycled according to Eq.[5.2]. The basic pulse sequence (a) is modified in (b) by the insertion of a $(\pi/4)_z$ pulse, together with a 90° clockwise shift of the receiver reference phase, in order to determine the sign of the double quantum frequency.

$$(\pi/2)_x - (\pi/2)_{-x} - (\pi/4)_{-y} - (\pi/2)_x - (\pi/4)_{-y} - (\pi/2)_x. \qquad [5.8]$$

In each sequence (a) and (b) of Fig.5.11, phase cycling according to Eq.[5.2] is used to cancel unwanted signal components from isolated ^{13}C spins, so that before the quadrature components are combined, they contain negligible contributions from molecules with isolated ^{13}C spins. The combination then results in a series of clean satellite spectra for different values of t_1, which are modulated in phase. As the sensitivity of the method is a critical problem, the acquisition time along the t_2-axis has to be at least 1.5 times the decay constant $T_2^{\star}$ of the signal. This generally corresponds to a large number of data points (>2000) along the t_2-axis, and therefore limits the allowed number of increments for t_1 to less than 100, giving a rather poor resolution in the F_1 dimension. Since signals in the F_1 dimension are usually well dispersed, this is no crucial problem.

As an illustration of the method the carbon-13 satellites of the carbon-13 spectrum of sucrose (Fig.5.12) were investigated. The Varian XL 200 spectrometer was operating at 50 MHz with a 16mm sample at 70^oC, and the total experimental time was 11 hours. The two-dimensional spectrum $S(\omega_1,\omega_2)$ is displayed in the form of an intensity contour plot. The F_2 dimension corresponds

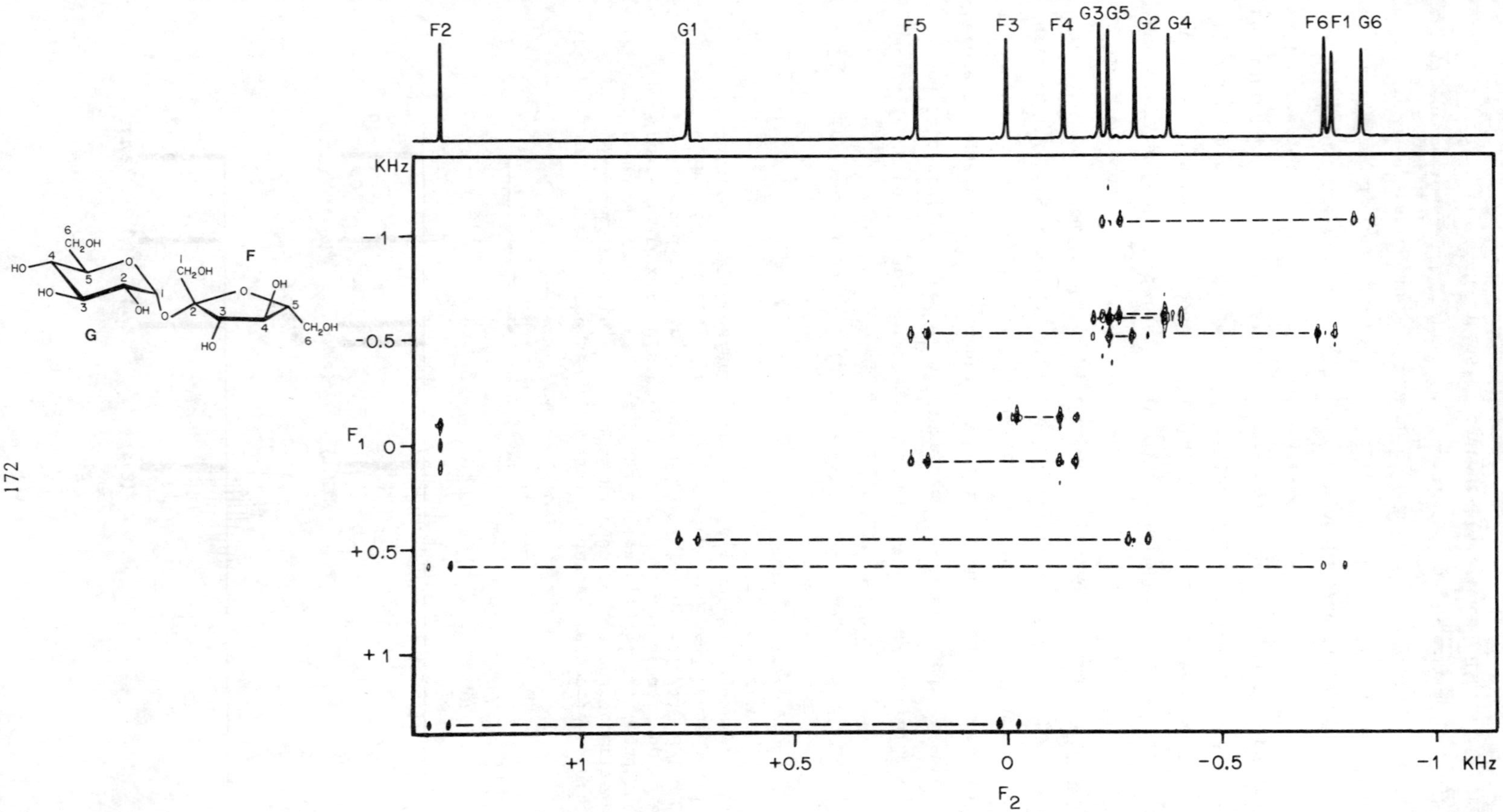

Fig.5.12 The two-dimensional carbon-13 spectrum of sucrose obtained as described in the text. The AX- or AB-type satellite spectra are the four-line patterns joined by broken lines; their centres of gravity lie on a line with $\Delta F_1/\Delta F_2 = 2$. The conventional carbon-13 spectrum running along the top of the diagram has been assigned by noting which resonances have a direct carbon-carbon spin coupling. The double quantum frequencies appear in the F_1 dimension. From ref.(13).

to the conventional carbon-13 spectrum except that the strong lines from isolated nuclei are suppressed, leaving only the weak satellite signals. Since the proton resonances are broad-band decoupled throughout, these are four-line spectra of the AX or AB type. Ten such spectra are visible in Fig. 5.12, marked by the broken lines; four of them show the characteristic AB intensity pattern. The F_1 dimension separates these spectra according to their individual double quantum frequencies, thus identifying them unequivocally. Since the double quantum frequency is equal to the sum of the appropriate carbon-13 shifts, the centres of gravity of all AX or AB spectra lie on a diagonal with $\Delta F_1/\Delta F_2 = 2$. This additional constraint could be useful for identifying artifacts due to incompletely suppressed signals from isolated carbon-13 nuclei.

The resonance of the quaternary carbon of sucrose (F2) was identified by its long spin-lattice relaxation time.

Inspection of Fig.5.12 reveals that this resonance is directly coupled to two others, which may then be labelled F1 and F3. F1 shows no other direct coupling, but F3 is clearly coupled through a single bond to another resonance, now labelled F4, and hence to F5 and F6. None of the glucose ring (G) resonances are involved in this coupling scheme. However, the three CH_2 sites are readily identified by their multiplicity and since F1 and F6 have been assigned to the fructose ring, G6 can be identified. The chain of linkages G6-G5-G4-G3-G2-G1 may then be deduced from Fig.5.12 by inspection; the fact that three AB quartets fall close together presents no particular difficulty if a larger plot of that region is made. The assignment agrees with that of Pfeffer et al. (17).

The double quantum technique provides a simple but powerful procedure for determing the connectivity of the carbon atoms in an organic molecule. (In principle C-N-C or C-O-C linkages could be bridged by searching out the long range C-C couplings by an extension of the method, using longer delay periods τ.) For materials with natural isotopic abundance the sensitivity is necessarily low, but the technique has the advantage of great simplicity because it examines isolated pairs of carbon-13 spins, and thus builds up a picture of the carbon skeleton in very simple steps.

A better way to distinguish between positive and negative modulation frequencies has recently be proposed by Mareci and Freeman (18). They have demonstrated that if the flip angle of the final pulse in the scheme of Fig. 5. 11a is set to 130° or 45°, phase modulation will occur and sensitivity will be enhanced by 30%.

References

(1) P.E. Hansen, Org. Magn. Reson. 11, 215 (1978).

(2) V. Wray, L. Ernst, T. Lund and H.J. Jakobsen, J. Magn. Reson. 40, 55 (1980).

(3) A. Bax, R. Freeman and S.P. Kempsell, J. Am. Chem. Soc. 102, 4849 (1980).

(4) D.I. Hoult and R.E. Richards, Proc. Roy. Soc. London A.344, 311 (1975).

(5) T.A. Frenkiel, Internal Report (1981).

(6) R. Freeman, S.P. Kempsell and M.H. Levitt, J. Magn. Reson. 38, 453 (1980).

(7) J.S. Waugh, J. Mol. Spectrosc. 35, 298 (1970).

(8) A. Bax and R. Freeman, J. Magn. Reson. 41, 507 (1980).

(9) W.P. Aue, E. Bartholdi and R.R. Ernst, J. Chem. Phys. 64, 2229 (1976).

(10) G. Bodenhausen, R. Freeman, G.A. Morris and D.L. Turner, J. Magn. Reson. 31, 75 (1978).

(11) A. Bax, S.P. Kempsell and R. Freeman, J. Magn. Reson. 41, 349 (1980).

(12) A. Bax, R. Freeman and T.A. Frenkiel, J. Am. Chem. Soc. 103, 2102 (1981).

(13) A. Bax, T.A. Frenkiel, R. Freeman and M.H. Levitt, J. Magn. Reson. 43, 478 (1981).

(14) R. Freeman, T.A. Frenkiel and M.H. Levitt, J. Magn. Reson. 44, 409 (1981).

(15) A. Bax, R. Freeman and G.A. Morris, J. Magn. Reson. 42, 164 (1981).

(16) A. Bax and G.A. Morris, J. Magn. Reson. 42, 501 (1981).

(17) P.E. Pfeffer, K.M. Valentine and F.W. Parrish, J. Am. Chem. Soc. 101, 1265 (1979).

(18) T.H. Mareci and R. Freeman, J. Magn. Reson. 48, 158 (1982).

6 PROBLEMS AND METHODS

6.1 Introduction

As mentioned before, two-dimensional NMR spectroscopy has turned out to be a very helpful tool in solving chemical problems. In many respects it can be seen as an alternative to the more familiar double résonance experiments. However, the sensitivity and ease of operation (after automatization of the spectrometer systems) of the new methods are generally better, and the practical limitations are often less severe. In order to be able to use two-dimensional spectroscopy, a modern NMR spectrometer with a flexible control system and a large amount of two-dimensional processing software is needed (see chapter 7).

Because of the abundance of two-dimensional methods and the large number of different applications for certain methods, in this chapter a survey will be given of possible problems and the corresponding two-dimensional methods to solve them. In Table 6.1 the information which can be obtained with various methods is tabulated. In the following paragraphs these topics will be discussed in more detail.

Only protons and carbon-13 nuclei are mentioned in this table. However, the methods can also be used in the studies of all other kinds of abundant or rare nuclei, although not many practical applications have yet been shown.

6.2 Methods for solving assignment problems

Assignment problems can be subdivided into two categories: spectra of abundant nuclei and spectra of rare nuclei. The commonest example of assignment problems in spectra of abundant nuclei is found in proton spin systems. Therefore in the next section attention will be focussed on assignment in proton spectra.

6.2.1 Assignment in proton spectra

The most common methods used in assigning proton spectra are the looking for identical multiplet splittings (matching of coupling constants) and selective homonuclear decoupling. The first method breaks down if couplings are not resolved, if many couplings have similar magnitude or if multiplets are overlapping in a very crowded spectrum. The second method, homonuclear decoupling, also breaks down in the case where resonances are broad compared with the J-splitting and in the case of strong coupling. There are several two-dimensional methods which can be useful alternatives or complementary

Table 6.1 A brief survey of problems and methods which can be used to solve them.

Assignment		section
proton spectra	homonuclear shift correlation spectroscopy	2.3, 2.4
	heteronuclear shift correlation spectroscopy	2.2
	cross-relaxation based shift correlation spectroscopy	2.4
	homonuclear J-spectroscopy	3.3
^{13}C spectra	heteronuclear J-spectroscopy	3.3
	heteronuclear shift correlation spectroscopy	2.2
	^{13}C - ^{13}C double quantum spectroscopy	5.5
Spectral parameters		
shift insensitive nuclei	heteronuclear shift correlation spectroscopy	2.2.5
homonuclear coupling constant	homonuclear J-spectroscopy	3.3
	homonuclear shift correlation spectroscopy	2.3
	the Inadequate experiment	5.2
	indirect J-spectroscopy	2.2.6
heteronuclear coupling constant	heteronuclear J-spectroscopy	3.2
	heteronuclear shift correlation spectroscopy	2.2
spin - spin relaxation rate	J-spectroscopy	3.2, 3.3
	zero quantum spectroscopy	4.5
	multiple quantum J-spectroscopy	4.5
exchange and cross-relaxation rate	exchange and cross-relaxation based shift correlation spectroscopy	2.4

tools in solving the assignment problem.

Homonuclear shift correlation spectroscopy

Homonuclear shift correlation spectroscopy as presented in paragraph 2.3 is a very simple and useful method to map out the connectivity pattern in a homonuclear coupled spin system. Depending on the sensitivity it is possible to show the presence of couplings which are up to ten times smaller than the natural line width. To get a first impression of the coupling pattern a good approach is to perform an experiment with short acquisition times ($\leqslant$ 0.2 sec.), along both time axes. For the mixing pulse α in the $\pi/2-t_1-\alpha-t_2$ sequence a width of $\pi/3$ or slightly larger can be chosen. A smaller mixing pulse will give relatively larger diagonal peaks but can make it possible to detect strongly coupled resonances unambiguously (section 2.3.3). In order to detect small couplings (< 2 Hz) longer acquisition times along both time axes are required. If this is not possible because of the limited size of data storage space available, delays can be introduced just before the evolution and detection periods (section 2.3.5). If strong overlap between the multiplets occurs, the broad-band decoupled variation of the latter experiment (section 2.3.6) can be useful.

Heteronuclear shift correlation spectroscopy

In the case where the assignment of the ^{13}C spectrum is known, and sensitivity permits ^{13}C detection, a simple and straight forward determination of the proton shifts is obtained using heteronuclear shift correlation spectroscopy as discussed in section 2.2, detecting the ^{13}C signal.

These indirectly obtained proton shifts can be used in the assignment of the individual resonances of the one-dimensional proton spectrum, which usually shows higher resolution.

Cross-relaxation based shift correlation spectroscopy

In the case of large molecules where homonuclear couplings are not resolved and the low sensitivity of ^{13}C does not permit the use of the previous method shift correlation spectroscopy based on cross-relaxation (section 2.4) can be used for solving the assignment problem. Practical examples of this application have been shown for the assignment in proton spectra (1). Especially at high field strength (> 7 Tesla) where longitudinal relaxation times tend to be long, while transverse relaxation times are short, this can be a very attractive method.

Homonuclear J-spectroscopy

Homonuclear J-spectroscopy (section 3.3) has initially been proposed as the ideal tool in the assignment of complicated spectra. However, the required weak coupling and the phase-twisted line shapes are problems in using the method. Its main use is probably limited to unraveling patterns of overlapping but individually well-resolved multiplets. Results are often rather disappointing in practical cases.

6.2.2 Assignment in spectra of rare nuclei

The most familiar rare nucleus is carbon-13, which will be used throughout this section as an example. However, most methods discussed are in principle also applicable to other kinds of rare nuclei.

The most common way of assigning ^{13}C spectra is the use of large tables of chemical shifts available for ^{13}C nuclei in a certain environment. Combined with the knowledge about substituent effects it is often possible to predict a ^{13}C chemical shift with an accuracy of a few ppm., allowing the direct assignment of recorded proton-decoupled ^{13}C spectra if the resonances are individually separated by a few ppm.

Remaining problems can sometimes be solved with the following methods:

Heteronuclear J-spectroscopy

The multiplicity n of a certain CH_n group can be obtained in a simple heteronuclear J-spectrum as described in section 3.2.4. New one-dimensional experiments, derived from two-dimensional spectroscopy offer the same possibilities (2,3,4).

Heteronuclear shift correlation spectroscopy

By performing the heteronuclear shift correlation experiment as described in section 2.2, the shifts of protons directly bonded to a carbon-13 are measured. If the assignment of the proton spectrum is known this can directly identify the ^{13}C. If the assignment of the proton or ^{13}C spectrum is partly known, the heteronuclear shift correlation spectrum will give additive information which can facilitate the assignment considerably.

^{13}C - ^{13}C Double quantum spectroscopy

By determing the ^{13}C - ^{13}C double quantum frequencies in a two-dimensional experiment as described in section 5.5 it is possible to determine the connectivity pattern and hence the assignment in the ^{13}C skeleton of a molecule in a straight forward and unambiguous way. The requirements for sensitivity and data storage space in this experiment are high.

6.3 Methods for extracting spectral parameters

Two-dimensional spectroscopy can be important for the determination of all kinds of spectral parameters like coupling constants, chemical shifts, relaxation rates, cross-relaxation rates and exchange rates. Below, the practical relevance of the different methods for measuring these parameters will be briefly discussed.

6.3.1 Shifts of nuclei with low magnetogyric ratio

As explained in section 2.2.5 the indirect detection in a heteronuclear shift correlation experiment of nuclei with a low magnetogyric ratio can be advantageous because of sensitivity reasons. Probably the most important application in this case is the determination of ^{15}N shifts, where in principle a sensitivity gain of a few orders of magnitude is possible. However, this technique has not yet been fully developed due to experimental difficulties (section 2.2.5).

6.3.2 Homonuclear coupling constants

In some cases homonuclear coupling constants cannot be extracted easily from a conventional one-dimensional spectrum. This is for instance the case in crowded spectral regions, in the case of a nearby much larger signal or just because of poor resolution.

Homonuclear J-spectroscopy

The two main advantages in using homonuclear J-spectroscopy in determing homonuclear coupling constants are the removal of overlap between the different multiplets, and the natural line widths in multiplet cross-sections (section 3.3). This latter effect is often not very important, provided that the field has been carefully shimmed. Only in the case of carefully degassed samples of small organic molecules can an improvement over line-narrowing by the use of digital filtering in one-dimensional spectroscopy be expected. In that case it is recommended to use the multiple refocussing technique as described in section 3.5.1.

Homonuclear shift correlation spectroscopy

If one is only interested in whether or not a (small) coupling between two nuclei does exist, the homonuclear shift correlation based on scalar interaction as discussed in section 2.3.5 is a very sensitive method to get the answer.

The Inadequate experiment

If coupled resonances are obscured because of the presence of a nearby much larger signal, as it is the case in natural abundance carbon-13 spectroscopy, the double quantum transfer Inadequate experiment, described in section 5.2, can be used to eliminate the large uncoupled resonance.

Indirect J-spectroscopy

Proton-proton coupling constants can be detected indirectly, e.g. via the carbon-13 nuclei (section 2.2.6) in a heteronuclear shift correlation experiment from which the proton shifts are removed. This offers the possibility to separate multiplets of protons which have exactly the same chemical shift, provided that the shifts of the directly coupled carbon-13 nuclei are different. The requirement of weak coupling among the protons remains in using this experiment.

6.3.3 Heteronuclear coupling constants

Heteronuclear J-spectroscopy

Sometimes the need exists to determine direct heteronuclear coupling constants with high accuracy, e.g. to detect the effect of a change in conformation or of a substituent. Since overlap in proton-coupled carbon-13 spectra is a commonly encountered problem, heteronuclear J-spectroscopy (section 3.2) can be fruitfully used.

Because of the natural line widths obtained with most methods, accurate results can be expected. Heteronuclear J-spectroscopy can also be used for the measurement of long range couplings. However, since long range couplings are often unresolved because of short transverse relaxation times and the fine structure of the multiplet is often very complex and the sensitivity in this application will be rather poor, often no satisfying results can be obtained. To avoid confusion due to strong coupling effects it is recommended to use the gated decoupler experiment with decoupled acquisition (section 3.2.2).

Heteronuclear shift correlation spectroscopy

The presence of a long range coupling can be shown, in analogy with homonuclear couplings, by performing a heteronuclear shift correlation experiment with proton-coupled evolution and detection. Nevertheless, this experiment remains rather difficult because of the poor sensitivity due to the fact that the spectrum is in the coupled mode along both frequency axes.

6.3.4 Spin-lattice relaxation

Aue et al. (5) pointed out that axial peaks in a homonuclear shift correlation spectrum contain information about the longitudinal relaxation rates. In practice, however, this does not appear to be very useful.

A nice example of the application of 2D spectroscopy for the determination of longitudinal relaxation times in complex overlapping proton spectra has been given by Avent and Freeman (6). An experiment proposed by Morris (7), using the Inept sequence (8) offers the same possibilities, but in a much simpler one-dimensional fashion.

6.3.5 Spin-spin relaxation

Two-dimensional spectroscopy can be useful in the determination of transverse relaxation rates. Not only single quantum relaxation rates but also zero- and multiple quantum relaxation rates are accessible.

J-spectroscopy

Since in principle natural line widths can be obtained with J-spectroscopy (sections 3.2 and 3.3), these widths can be used directly for obtaining a value for the relaxation time if the line has a Lorentzian shape.

Zero quantum spectroscopy

Multiple quantum spectroscopy gives natural line widths for homonuclear zero quantum transitions, enabling the determination of the relaxation rates directly and with high accuracy (section 4.5).

Multiple quantum J-spectroscopy

Multiple quantum J-spectroscopy as discussed in section 4.5 makes it possible to determine the relaxation rates of all detectable multiple quantum transitions via line width determination.

6.3.6 Exchange and cross-relaxation rates

Exchange and cross-relaxation spectroscopy as developed by the groups of Jeener and Ernst (9,1) and briefly presented in section 2.4, is a powerful tool to show the presence of cross-relaxation and exchange. The exact measurement of the rates using these methods is rather difficult. An estimate for the order of magnitude can be obtained by varying the length of the mixing period in consecutive experiments or by comparing the amplitudes of diagonal and cross peaks.

References

(1) A. Kumar, R.R. Ernst and K. Wüthrich, Biochem. Biophys. Res. Commun. 95, 1 (1980).

(2) R. Freeman and M.H. Levitt, J. Magn. Reson. 39, 533 (1980).

(3) D.P. Burum and R.R. Ernst, J. Magn. Reson. 39, 163 (1980).

(4) D.M. Doddrell and D.T. Pegg, J. Amer. Chem. Soc. 102, 6388 (1980).

(5) W.P. Aue, E. Bartholdi and R.R. Ernst, J. Chem. Phys. 64, 2229 (1976).

(6) A.G. Avent and R. Freeman, J. Magn. Reson. 39, 169 (1980).

(7) G.A. Morris, J. Magn. Reson. 41, 185 (1980).

(8) G.A. Morris and R. Freeman, J. Amer. Chem. Soc. 101, 760 (1979).

(9) J. Jeener, B.H. Meier, P. Bachmann and R.R. Ernst, J. Chem. Phys. 71, 4546 (1979).

7 SPECTROMETER- AND COMPUTER REQUIREMENTS

7.1 Introduction

In this chapter it will be briefly discussed in a qualitative way which spectrometer capabilities are more critical for two-dimensional experiments than for the common one-dimensional experiments. It will appear that a modern NMR spectrometer needs only a relatively small extension of software and possibly upgrading of plotting and background storage facilities. The main hardware requirements for two-dimensional spectroscopy also apply to performing sophisticated one-dimensional experiments. The main purpose of this chapter is to give an impression to which extent a modern NMR spectrometer has to be modified in order to be suitable for 2D spectroscopy.

7.2 Hardware

In this section the analog and the control hardware will be discussed. Many of the normal requirements for the analog hardware of NMR spectrometers are quite critical for 2D spectroscopy. These requirements are:

(a) Field frequency lock system

A change in field frequency ratio during the 2D experiment can give rise to t_1-noise (section 1.8), line shape distortions, and loss of sensitivity. Since 2D experiments usually last several hours, optimization of the lock system is important.

(b) Constant field homogeneity

A change in homogeneity will again result in t_1-noise, line shape distortion, and possibly extra spurious resonance lines.

(c) Spinning sidebands

The effect of spinning sidebands in two-dimensional spectroscopy (1) is generally even more confusing than in a one-dimensional spectrum. Since the range of different intensities in 2D spectra can be large, it can be difficult to distinguish between real peaks and spinning sidebands purely on the basis of their magnitude.

(d) Homogeneous r.f.fields

Inhomogeneous radiofrequency fields generated by pulses can cause artefacts in the 2D spectrum. For the observing channel on modern NMR spectrometers, the r.f.field homogeneity is usually sufficient, however, also for the heteronuclear channel a homogeneous r.f.field is needed. Demands for r.f.field homogeneity can be weakened by the use of composite pulses (2).

(e) Homogeneity destroy pulse

A homogeneity destroy pulse (homospoil) is a handy and sometimes required tool in 2D NMR experiments (sections 2.3.2, 2.4, 4.4.2). If this homospoil pulse does affect the field frequency lock system, or if the timing of the pulse is bad, t_1-noise, artefacts and loss of sensitivity can occur.

The demands for control hardware of course depend largely on the available control software (3). Apart from that a number of requirements can be named which the combination of control hard- and software must be capable to fulfil:

(a) Pulsing on the hetero-channel

For many experiments it must be possible to apply pulses to the heteronuclei. Not only should the length of this pulse be selectable within a few percent of the desired value, but also the phase should be settable to an integer multiple of 90°.

(b) Accuracy of timing

The timer of the spectrometer should be capable of generating intervals between pulses with any length between a few microsec. and several seconds with a high absolute accuracy of a few microsec. If the accuracy of timing is insufficient, t_1-noise and extra lines can appear in the 2D spectrum.

(c) Flexible acquisition settings

In order to economize on the size of the data matrix it is necessary to be able to select a spectral width and an acquisition time close to the desired values. Of course, a suitable analog filter must be available for any bandwidth chosen.

(d) Transmitter frequencies

In order to economize on the size of the data matrix it should also be possible to set the observing transmitter and the heteronuclear transmitter close to the centre of the spectrum.

(e) Pulse calibration

If the flip angle of r.f.pulses cannot be calibrated accurately to $\pi/2$ or to π radians artefacts in the 2D spectrum can occur.

7.3 Software

7.3.1 Control and acquisition software

Of course, the specifications for control software depend on the capabilities of the control hardware (3). Generally, however, it can be stated that the combination of control hard- and software should be capable of creating any pulse sequence with a number of up to some 15 pulses in a flexible fashion, i.e. the time intervals between these pulses should be freely selectable, the pulses can be applied to the observe and/or hetero channel, and the phases of the pulses are selectable. In order to perform a two-dimensional experiment it is necessary that the delays between pulses can be incremented (or decremented) in successive experiments by a selectable length, while the corresponding data are stored separately on a background storage device. The possibility should exist to perform a certain experiment (for a single t_1-value) an arbitrary number of times, while phases of the pulses and receiver are programmable for these repetitions. In this case the data should be directly averaged to the previously acquired data for the same length of t_1. During all intervals between pulses it should be selectable whether or not to apply heteronuclear decoupling.

7.3.2 Processing software

As mentioned in section 1.8, the extra amount of data processing software for performing 2D experiments is rather limited. In fact, only an extra matrix transposition routine is needed. However, a few extra demands are made on the individual processing programs to facilitate the actual operation.

(a) The original data should not be overwritten by the processed data, in order to allow reprocessing with different parameters.

(b) The processing should be fast. Since the data matrices are usually fairly large, processing should be fast in order to limit the amount of processing time. Not only the computer calculation speed, but also a minimum number of disc accesses is of importance in this case.

(c) It should be possible to define all processing operations in advance (queuing) to avoid a cumbersome and time-consuming operator interaction.

7.3.3 Display software

The display of two-dimensionalspectra is usually a time-consuming process because of the large amount of data involved and because of the complicated arithmetic needed for white-wash or contour calculation procedures, and even more the usually fairly slow mechanical plotting device.

In order to minimize plotting time it is necessary to get a (crude) pre-view of the two-dimensional spectrum on a fast display unit (e.g. storage scope or C.R.T. terminal) as it would appear on the plotter, with the option of changing the display parameters. This avoids the making of useless plottings. A proper contour plotting program is for many 2D experiments absolutely necessary, while also a stacked-trace plot program can in some cases be very useful. Further on, it is nice if the selection of cross-sections which contain resonances is automated. This is for instance useful in heteronuclear shift correlation (section 2.2) and in homonuclear J-spectroscopy (4) (section 3.3).

7.4 Computer and peripherals

The computer specifications required for being able to perform two-dimensional spectroscopy, do not differ much in principle from the requirements which are made anyway on a computer controlling a modern HR - NMR spectrometer. It must be capable of controlling the spectrometer and acquiring, processing and storing the data. The computer-spectrometer system must be capable of generating the pulse sequences of the various 2D experiments and acquiring the data for a large number of different lengths of the evolution period. The major difference with respect to the computer requirements between one- and two-dimensional spectroscopy is caused by the very large amount of data usually involved. Therefore a large background storage device (magnetic disc) is needed with at least storage space for a number of data points of the order of 10^6. In order to limit the processing time of the large amount of data it is preferable, although not a prerequisite, if the following features are available:

(a) fast instruction execution time, possibly by using an array processor,
(b) fast access to the background storage,
(c) large computer memory to minimize the number of accesses to the background storage device.

With respect to the peripherals, a fast plotting device is needed, which is capable of making contour- and stacked-trace plots (section 1.6.2). A fast display unit for displaying the 2D spectrum before it is plotted is attractive in order to limit the number of replottings of the final spectrum. For interactive phasing of the two-dimensional spectrum (5) a fast display unit is essential.

References

(1) G. Bodenhausen, S.P. Kempsell, R. Freeman and H.D. Hill, J. Magn. Reson. 35, 337 (1979).

(2) M.H. Levitt and R. Freeman, J. Magn. Reson. 43, 65 (1981).

(3) A.F. Mehlkopf, A. Bax, T.A. Tiggelman and J. Smidt, Bull. of Magn. Reson.,(to be published).

(4) K. Nagayama, P. Bachmann, K. Wüthrich and R.R. Ernst, J. Magn. Reson. 31, 133 (1978).

(5) M.H. Levitt and R. Freeman, J. Magn. Reson. 34, 675 (1979)

APPENDIX I

Example of a density matrix calculation

In this appendix an example will be given how the theory of section 1.2.3 can be used to calculate the behaviour of the spin system explicitly. The effect of the pulse sequence of Fig.I.1, used in section 5.5 on a homonuclear AX spin system in thermal equilibrium will be analyzed.

The rotation operators for a $(\pi/2)_x$, a $(\pi)_y$ and a $(\pi/2)_y$ pulse, as derived by Schäublin et al.(1), are given below.

$$\hat{R}_x^{-1}(\pi/2) = \tfrac{1}{2}\begin{bmatrix} 1 & -i & -i & -1 \\ -i & 1 & -1 & -i \\ -i & -1 & 1 & -i \\ -1 & -i & -i & 1 \end{bmatrix} ; \quad \hat{R}_x(\pi/2) = \tfrac{1}{2}\begin{bmatrix} 1 & i & i & -1 \\ i & 1 & -1 & i \\ i & -1 & 1 & i \\ -1 & i & i & 1 \end{bmatrix}$$

$$\hat{R}_y(\pi) = \hat{R}_y^{-1}(\pi) = \begin{bmatrix} 0 & 0 & 0 & 1 \\ 0 & 0 & -1 & 0 \\ 0 & -1 & 0 & 0 \\ 1 & 0 & 0 & 0 \end{bmatrix} \qquad [I.1]$$

$$\hat{R}_y^{-1}(\pi/2) = \tfrac{1}{2}\begin{bmatrix} 1 & 1 & 1 & 1 \\ -1 & 1 & -1 & 1 \\ -1 & -1 & 1 & 1 \\ 1 & -1 & -1 & 1 \end{bmatrix} ; \quad \hat{R}_y(\pi/2) = \tfrac{1}{2}\begin{bmatrix} 1 & -1 & -1 & 1 \\ 1 & 1 & -1 & 1 \\ 1 & -1 & 1 & -1 \\ 1 & 1 & 1 & 1 \end{bmatrix}$$

The notation used for the density matrix at certain times is indicated in Fig.I.1.

The energy level diagram of the AX system is sketched in Fig.I.2. The energies of the different levels are indicated too.

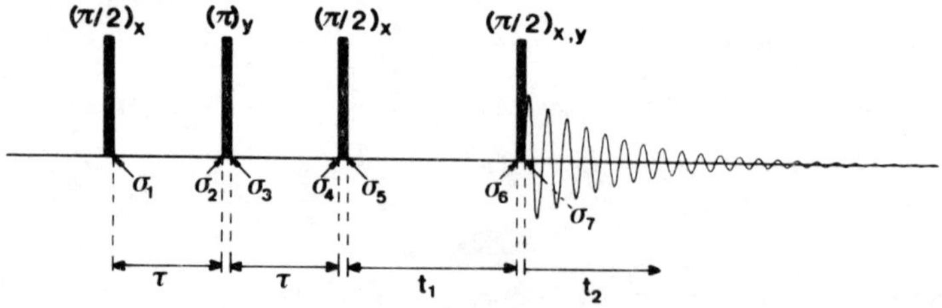

Fig.I.1 Pulse sequence of which the effect on an AX spin system is calculated.

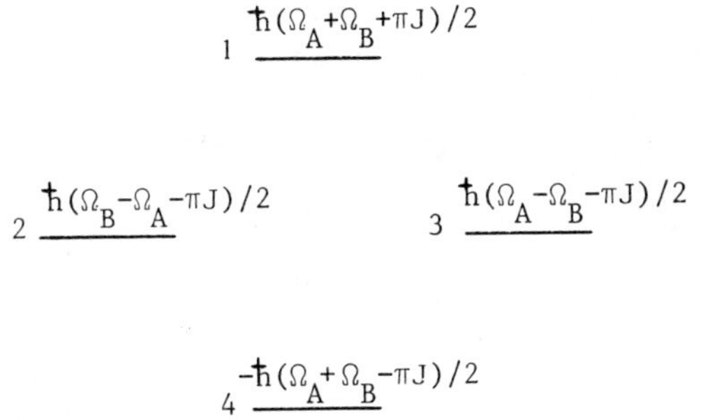

Fig.I.2 Energy level diagram and energies in an AX spin system.

The density matrix at thermal equilibrium is given by

$$\sigma_{eq} = \frac{1}{4}\begin{bmatrix} 1+p & 0 & 0 & 0 \\ 0 & 1 & 0 & 0 \\ 0 & 0 & 1 & 0 \\ 0 & 0 & 0 & 1-p \end{bmatrix} = \frac{1}{4}\,\mathbb{1} + p/4\begin{bmatrix} 1 & 0 & 0 & 0 \\ 0 & 0 & 0 & 0 \\ 0 & 0 & 0 & 0 \\ 0 & 0 & 0 & -1 \end{bmatrix}, \qquad [I.2]$$

where $p = \hbar\Omega_A/kT$. Ω_A is the angular Larmor frequency of spin A, and $\Omega_A/\Omega_B \approx 1$. Since all rotation operators commute with the unity operator $\mathbb{1}$, this part of the density matrix remains unchanged during the experiment and will be omitted in the expressions below.

One finds:

$$\sigma_1 = \hat{R}_x^{-1}(\pi/2)\sigma_{eq}\hat{R}_x(\pi/2) = p/8i\begin{bmatrix} 0 & 1 & 1 & 0 \\ -1 & 0 & 0 & 1 \\ -1 & 0 & 0 & 1 \\ 0 & -1 & -1 & 0 \end{bmatrix} = p/4\ \hat{F}_y. \qquad [I.3]$$

σ_2 follows using Eq. [1.8] :

$$\sigma_2 = p/8i\begin{bmatrix} 0 & e^{i\omega_{12}\tau} & e^{i\omega_{13}\tau} & 0 \\ -e^{i\omega_{21}\tau} & 0 & 0 & e^{i\omega_{24}\tau} \\ -e^{i\omega_{31}\tau} & 0 & 0 & e^{i\omega_{34}\tau} \\ 0 & -e^{i\omega_{42}\tau} & -e^{i\omega_{43}\tau} & 0 \end{bmatrix} \qquad [I.4]$$

with $\omega_{mn} = (E_n - E_m)/\hbar$, and the values for E_n given in Fig.I.2, e.g. $\omega_{12} = -\Omega_A - \pi J$, $\omega_{21} = \Omega_A + \pi J$.

For σ_3 one finds:

$$\sigma_3 = \hat{R}_y^{-1}(\pi)\sigma_2\hat{R}_y(\pi) = p/8i \begin{bmatrix} 0 & e^{i\omega_{43}\tau} & e^{i\omega_{42}\tau} & 0 \\ -e^{i\omega_{34}\tau} & 0 & 0 & e^{i\omega_{31}\tau} \\ -e^{i\omega_{24}\tau} & 0 & 0 & e^{i\omega_{21}\tau} \\ 0 & -e^{i\omega_{13}\tau} & -e^{i\omega_{12}\tau} & 0 \end{bmatrix} . \quad [I.5]$$

Using Eq. [1.8] again gives

$$\sigma_4 = p/8i \begin{bmatrix} 0 & a & a & 0 \\ -a^{\times} & 0 & 0 & a^{\times} \\ -a^{\times} & 0 & 0 & a^{\times} \\ 0 & -a & -a & 0 \end{bmatrix} \quad [I.6]$$

with $a = \exp(-i2\pi J\tau)$.

Using the arguments given in section 1.2.3 (Eq. [1.22]) about an average Hamiltonian $\bar{\bar{H}}$, would have given σ_4 directly from σ_1.

The second $(\pi/2)_x$ pulse creates the double quantum coherence, as can be seen in the form of σ_5, which contains elements σ_{14} and σ_{41}, not equal to zero:

$$\sigma_5 = \hat{R}_x^{-1}(\pi/2)\sigma_4\hat{R}_x(\pi/2) = p/4 \begin{bmatrix} -\cos(2\pi J\tau) & 0 & 0 & i\sin(2\pi J\tau) \\ 0 & 0 & 0 & 0 \\ 0 & 0 & 0 & 0 \\ -i\sin(2\pi J\tau) & 0 & 0 & \cos(2\pi J\tau) \end{bmatrix} . \quad [I.7]$$

Assuming that τ has been chosen to be equal to $(4J)^{-1}$, only double quantum coherence will be present during the time t_1, and σ_5 simplifies to

$$\sigma_5 = p/4 \begin{bmatrix} 0 & 0 & 0 & i \\ 0 & 0 & 0 & 0 \\ 0 & 0 & 0 & 0 \\ -i & 0 & 0 & 0 \end{bmatrix} . \quad [I.8]$$

Using Eq. [1.8] again, one finds

$$\sigma_6 = p/4 \begin{bmatrix} 0 & 0 & 0 & ie^{i\omega_{14}t_1} \\ 0 & 0 & 0 & 0 \\ 0 & 0 & 0 & 0 \\ -ie^{i\omega_{41}t_1} & 0 & 0 & 0 \end{bmatrix} . \quad [I.9]$$

In the case where the last $\pi/2$ pulse is applied along the x-axis one finds for σ_7:

$$\sigma_7 = \hat{R}_x^{-1}(\pi/2)\sigma_6\hat{R}_x(\pi/2) = p/8\begin{bmatrix} s & c & c & -s \\ c & -s & -s & -c \\ c & -s & -s & -c \\ -s & -c & -c & s \end{bmatrix} \qquad [I.10]$$

with $s = \sin(\omega_{14}t_1)$ and $c = \cos(\omega_{24}t_1)$.

In the case where the last $\pi/2$ pulse is applied along the y-axis one finds for σ_7:

$$\sigma_7 = \hat{R}_y^{-1}(\pi/2)\sigma_6\hat{R}_y(\pi/2) = p/8\begin{bmatrix} -s & -ic & -ic & -s \\ ic & s & s & ic \\ ic & s & s & ic \\ -s & -ic & -ic & -s \end{bmatrix}. \qquad [I.11]$$

For the magnetization component $M_{tr_{12}}$ which corresponds with coherence between the levels 1 and 2 one finds, using Eq. [1.26],

$$M_{tr12} = C\ Tr(\hat{F}_{x12} + i\hat{F}_{y12})\sigma_7. \qquad [I.12]$$

In the case of a final $(\pi/2)_x$ pulse it follows from Eqs. [I.10] and [I.12]:

$$M_{tr12} = pC/8\ \cos(\omega_{14}t_1). \qquad [I.13]$$

In the case of a final $(\pi/2)_y$ pulse it follows from Eqs. [I.11] and [I.12]:

$$M_{tr12} = -ipC/8\ \cos(\omega_{14}t_1). \qquad [I.14]$$

Note that Eq. [I.13] represents a signal along the x-axis while Eq. [I.14] is a signal along the -y-axis. This agrees with the results given in Table 5.1. During the time t_2 the magnetization component will precess with angular frequency ω_{12} and give a signal

$$s(t_1,t_2) = pC/8\ \cos(\omega_{14}t_1)\exp(i\omega_{12}t_2) \qquad [I.15a]$$

in the case of a final $(\pi/2)_x$ pulse, and

$$s(t_1,t_2) = pC/8\ \cos(\omega_{14}t_1)\exp\{i(\omega_{12}t_2 - \pi/2)\} \qquad [I.15b]$$

in the case of a final $(\pi/2)_y$ pulse.

Reference

(1) S. Schäublin, A. Höhener and R.R. Ernst, J. Magn. Reson. 13, 196 (1974).

APPENDIX II

Magnetization transfer in $^{13}CH_2$ and $^{13}CH_3$ groups

In this appendix the effect of the heteronuclear shift correlation experiment (Fig.2.2) on the signals originating from a $^{13}CH_2$ group will be analyzed. After that the effect of heteronuclear decoupling as described in section 2.2.2 will be treated. The derivations for a $^{13}CH_3$ group are completely analogous and only the final results will be given.

The energy level diagram of an isolated $^{13}CH_2$ with equivalent protons is given in Fig.II.1. The corresponding wave functions and energies are given in Table II.1.

Table II.1 Wave functions and energies corresponding to the different energy levels in the diagram II.1. The first spin in the notation used denotes the ^{13}C-nucleus with chemical shift frequency Ω_2; Ω_1 denotes the proton chemical shift frequency.

Energy level	Wave function	Energy
1	$\beta\beta\beta$	$\frac{1}{2}\hbar(\Omega_2 + 2\Omega_1 + 2\pi J + \pi J_{HH})$
2	$\alpha\beta\beta$	$\frac{1}{2}\hbar(-\Omega_2 + 2\Omega_1 - 2\pi J + \pi J_{HH})$
3	$\frac{1}{2}\sqrt{2}\ \beta(\alpha\beta + \beta\alpha)$	$\frac{1}{2}\hbar(\Omega_2 + \pi J_{HH})$
4	$\frac{1}{2}\sqrt{2}\ \alpha(\alpha\beta + \beta\alpha)$	$\frac{1}{2}\hbar(-\Omega_2 + \pi J_{HH})$
5	$\beta\alpha\alpha$	$\frac{1}{2}\hbar(\Omega_2 - 2\Omega_1 - 2\pi J + \pi J_{HH})$
6	$\alpha\alpha\alpha$	$\frac{1}{2}\hbar(-\Omega_2 - 2\Omega_1 - 2\pi J + \pi J_{HH})$
7	$\frac{1}{2}\sqrt{2}\ \beta(\alpha\beta - \beta\alpha)$	$\frac{1}{2}\hbar(\Omega_2 - 3\pi J_{HH})$
8	$\frac{1}{2}\sqrt{2}\ \alpha(\alpha\beta - \beta\alpha)$	$\frac{1}{2}\hbar(-\Omega_2 - 3\pi J_{HH})$

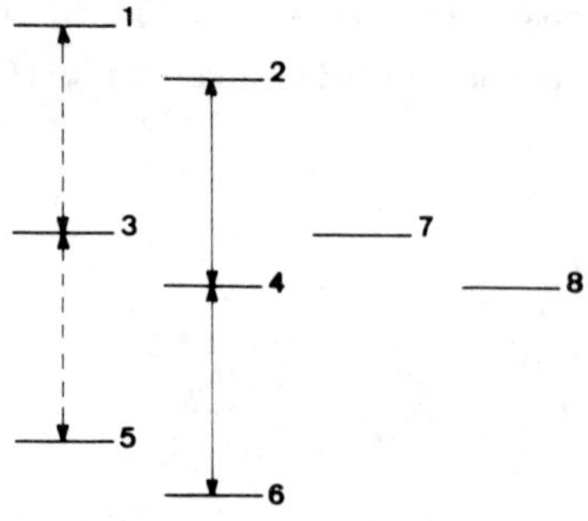

Fig.II.1 Energy level diagram of a $^{13}CH_2$ spin system. The corresponding energies and wave functions are given in Table II.1. The broken and drawn transitions correspond to the proton doublet lines with frequencies $\Omega_1+\pi J$ and $\Omega_1-\pi J$.

As expected, and as can be deduced from Table II.1, only two different proton resonances exist, at frequencies $\Omega_1 \pm \pi J$ if Ω_1 is the ^1H chemical shift frequency and J the heteronuclear coupling constant.

In Fig.II.1 the transitions with resonance frequency $\Omega_1 - \pi J$ and $\Omega_1 + \pi J$ are indicated by the drawn and broken lines respectively.

The effect of the pulse sequence of Fig.2.2 on the magnetization in the $^{13}CH_2$ spin system will now be explained. The first $(\pi/2)_x$ pulse applied to the protons will create two transverse magnetization vectors (Fig.2.3a), corresponding to the drawn and broken lines in Fig.II.1. The second $(\pi/2)_x$ pulse applied to the protons will create longitudinal ^1H magnetization components, analogous to Eq. [2.2] given by

$$M_{z+} = -M_o(^1H)\cos\{(\Omega_1 + \pi J)t_1\} \qquad [II.1a]$$

$$M_{z-} = -M_o(^1H)\cos\{(\Omega_1 - \pi J)t_1\} \;, \qquad [II.1b]$$

where M_{z+} and M_{z-} denote the longitudinal proton magnetization corresponding to the broken and drawn transitions in Fig.II.1 $M_o(^1H)$ is the thermal equilibrium magnetization of the protons corresponding to the drawn and to the broken transitions. Since the two proton pulses do not affect the ^{13}C spin the sums of the populations of the levels with the ^{13}C spin in the α-state (or β-state) remains unchanged during the pulse sequence applied to the proton. This gives the relations

$$P_1(t_1) + P_3(t_1) + P_5(t_1) = 3(1 - p) \qquad [II.2a]$$

$$P_2(t_1) + P_4(t_1) + P_6(t_1) = 3(1 + p) \qquad [II.2b]$$

using the thermal equilibrium populations as given in Table II.2. Equation [II.2] is the equivalent of Eq. [2.3] in section 2.2.1. It can be derived that the individual longitudinal magnetization components corresponding to each of the drawn (or broken) transitions are equal after the second proton pulse. This implies that

$$P_1(t_1) - P_3(t_1) = P_3(t_1) - P_5(t_1) \qquad [II.3a]$$

$$P_2(t_1) - P_4(t_1) = P_4(t_1) - P_6(t_1). \qquad [II.3b]$$

From Eqs. [II.1], [II.2] and [II.3] it then follows that the new populations after the second proton pulse are as given in Table II.2.

Table II.2 Populations of the different energy levels in Fig.II.1 at thermal equilibrium (P_n), and after a $(\pi/2)_x - t_1 - (\pi/2)_x$ pulse sequence applied to the protons $\{P_n(t_1)\}$. The constant p equals $\gamma_{13C}H_o\hbar/2kT$.

Level	P_n	$P_n(t_1)$
1	$1 - 9p$	$1 - p + 8p\cos\{(\Omega_1 + \pi J)t_1\}$
2	$1 - 7p$	$1 + p + 8p\cos\{(\Omega_1 - \pi J)t_1\}$
3,7	$1 - p$	$1 - p$
4,8	$1 + p$	$1 + p$
5	$1 + 7p$	$1 - p - 8p\cos\{(\Omega_1 + \pi J)t_1\}$
6	$1 + 9p$	$1 + p - 8p\cos\{(\Omega_1 - \pi J)t_1\}$

In thermal equilibrium the magnetization M_o corresponding to a single ^{13}C transition (e.g. 1-2) corresponds to a difference in population of 2p of the two levels involved. Using the new populations of Table II.2 one finds for the new longitudinal ^{13}C magnetizations:

$$M_{z12}(t_1) = M_o + 4M_o[\cos\{(\Omega_1 + \pi J)t_1\} - \cos\{(\Omega_1 - \pi J)t_1\}] \quad \text{[II.4a]}$$

$$M_{z34}(t_1) = M_{z78}(t_1) = M_o \quad \text{[II.4b]}$$

$$M_{z56}(t_1) = M_o - 4M_o[\cos\{(\Omega_1 + \pi J)t_1\} - \cos\{(\Omega_1 - \pi J)t_1\}]. \quad \text{[II.4c]}$$

A ^{13}C $\pi/2$ pulse, applied immediately after the second proton pulse, will create transverse magnetization which induces a signal given by

$$s(t_1,t_2) = M_{z12}(t_1)\exp\{i(\Omega_2 + 2\pi Jt_1)\} + 2M_o\exp(i\Omega_2 t_1) + M_{z56}(t_1) \times \times\exp\{i(\Omega_2 - 2\pi Jt_1)\}, \quad \text{[II.5]}$$

where Ω_2 denotes the ^{13}C chemical shift frequency.

From Eqs. [II.4] and [II.5] it follows that the peak positions in a two-dimensional spectrum are given by

$$(\omega_1,\omega_2) = (0,\Omega_2 - 2\pi J),\ (0,\Omega_2),\ (0,\Omega_2 + 2\pi J),\ (\Omega_1 + \pi J,\Omega_2 - 2\pi J),$$
$$(\Omega_1 + \pi J,\Omega_2 + 2\pi J),\ (\Omega_1 - \pi J,\Omega_2 - 2\pi J),\ (\Omega_1 - \pi J,\Omega_2 + 2\pi J).$$

The relative intensities of these peaks are 1:2:1:4:-4:-4:4.

If broad-band proton decoupling were started at an arbitrary time Δ_2 after the ^{13}C $\pi/2$ pulse one finds, analogous to Eq. [2.7], for the detected signals:

$$s(t_1,t_2) = 2M_o\{1 + \cos(2\pi J\Delta_2)\}\exp\{i\Omega_2(t_2 + \Delta_2)\} - 8iM_o\sin(2\pi J\Delta_2) \times \times[\cos\{(\Omega_1 + \pi J)t_1\} - \cos\{(\Omega_1 - \pi J)t_1\}] \times \exp\{i\Omega_2(t_2 + \Delta_2)\}. \quad \text{[II.6]}$$

For a $^{13}CH_3$ group it can be derived in a similar way that the detected signal is given by

$$s(t_1,t_2) = 2M_o\{\cos(3\pi J\Delta_2) + 3\cos(\pi J\Delta_2)\}\exp i\Omega_2(t_2 + \Delta_2) - \\ - 12iM_o\{\sin(\pi J\Delta_2) + \sin(3\pi J\Delta_2)\}[\cos\{(\Omega_1 + \pi J)t_1\} - \\ - \cos\{(\Omega_1 - \pi J)t_1\}]\exp\{i\Omega_2(t_2 + \Delta_2)\}. \qquad [II.7]$$

The amplitudes of the signals that are modulated as a function of t_1 is proportional to $\sin(2\pi J\Delta_2)$ and $\{\sin(\pi J\Delta_2) + \sin(3\pi J\Delta_2)\}$ in the case of $^{13}CH_2$ and $^{13}CH_3$ groups respectively. A graphical presentation of this behaviour is given in Fig.2.9.

Comparing Eqs. [II.6] and [II.7] with Eq. [2.7] shows that the modulation of the signals as a function of t_1 is identical in the cases of a ^{13}CH, $^{13}CH_2$ and a $^{13}CH_3$ group. Of course, this should be expected, since the proton spectrum is in all three cases a doublet with splitting J.

^{13}C Decoupling during the evolution period can thus be performed in exactly the same way as described in section 2.2.2, with the optimum value of Δ_1 again equal to $(2J)^{-1}$.

APPENDIX III

The effect of a single selective pulse

In this appendix the effect of a single line-selective pulse (see footnote section 1.2.3), applied to a coupled spin system, will be analyzed.

First an uncomplete and very simple approach to the problem will be made not taking into account the phase of the selective pulse; the second part of this appendix deals with the mathematical calculation, using fictitious spin-½ operators.

Consider a set of isolated spin-½ nuclei in a magnetic field. The two eigenstates are $|\alpha\rangle$ and $|\beta\rangle$. The populations of the two levels are written as P_α and P_β.

The longitudinal magnetization M_o in the spin system is given by

$$M_o = N\gamma\hbar(P_\alpha - P_\beta), \qquad \text{[III.1]}$$

where N is the number of spins in the sample. Suppose for reasons of convenience that all spins are in state $|\alpha\rangle$, i.e. $M_o = N\gamma\hbar$. If a pulse with flip angle θ is applied along the x-axis, this will rotate the magnetization over an angle θ about the x-axis. The new longitudinal magnetization is then equal to $N\gamma\hbar\cos\theta$. The new occupations of the energy levels are then given by

$$\begin{aligned} P_\alpha &= \cos^2(\theta/2) \\ P_\beta &= \sin^2(\theta/2). \end{aligned} \qquad \text{[III.2]}$$

In the quantum mechanical formalism the wave function Ψ of a certain spin is written as $\Psi = c_\alpha|\alpha\rangle + c_\beta|\beta\rangle$.

The population of the energy level corresponding with state α is equal to $\overline{|c_\alpha|^2}$ Hence, the effect of a pulse with flip angle θ on a spin in state $|\alpha\rangle$ $(\Psi = 1|\alpha\rangle + 0|\beta\rangle)$ is to alter the wave function according to ★

$$\Psi' = (P_\alpha)^{\frac{1}{2}}|\alpha\rangle + (P_\beta)^{\frac{1}{2}}|\beta\rangle = \cos(\theta/2)|\alpha\rangle + \sin(\theta/2)|\beta\rangle. \qquad \text{[III.3]}$$

This is the central equation in the simple approach which can be used to visualize the effect of a selective pulse in a coupled spin system.

Consider for example a part of an energy level diagram of a coupled spin system, as sketched in Fig.III.1. Assume that a coherence is present be-

---

★ A more detailed quantum mechanical analysis shows that the phase of state β is generally not zero as assumed in Eq. [III.3], but depends on the phase of the pulse.

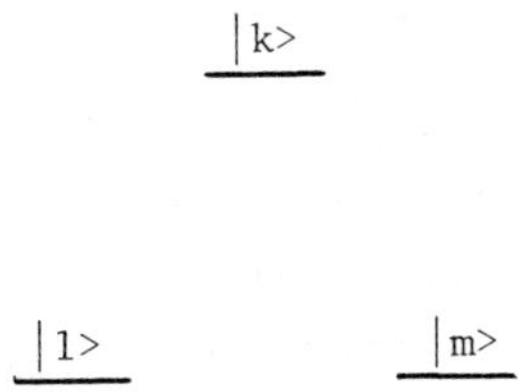

Fig.III.1 Part of the energy level diagram of a coupled spin system.

tween the states k and l. This coherence is described by density matrix element σ_{kl} (and its complex conjugated σ_{lk}):

$$\sigma_{kl} = \overline{c_l^{\times} c_k} \ . \qquad \text{[III.4]}$$

Change in coherence kl

Due to a selective pulse with flip angle θ on transition km {a θ(km) pulse}, the magnitude of c_k changes according to Eq.[III.3] by a factor $\sin(\theta/2)$, and using Eq.[III.4] this implies that a factor $\cos(\theta/2)$ of coherence between the states |k> and |l> is left.

Transfer to coherence lm

Suppose for reasons of convenience that the occupation of level m before the selective θ(km) pulse equals zero (i.e. $c_m = 0$). Immediately after the θ(km) pulse the coefficient c_m of an individual spin is related to c_k by Eq.[III.3]:

$$c_m = c_k \sin(\theta/2). \qquad \text{[III.5]}$$

Therefore, if an ensemble averaged phase coherence $\sigma_{kl} = \overline{c_l^{\times} c_k}$ is present before the θ(km) pulse, a phase coherence $\sigma'_{ml} = \overline{c_l^{\times} c_m}$ will be created, related to σ_{kl} by

$$\sigma'_{ml} = \sin(\theta/2)\sigma_{kl}, \qquad \text{[III.6]}$$

an amount proportional to $\sin(\theta/2)$ is transferred from coherence between the levels k and l to coherence between the states |l> and |m>.

Fictitious spin-½ operators

A more exact treatment of the effect of a selective pulse can be obtained if the fictitious spin-½ arithmetic is used. First a description of the formalism will be given and then the relation with physics will be shown.

The definition of fictitious spin-½ operators as given by Wokaun and Ernst (1,2) will be used here. For any coherence between two eigenstates $|r\rangle$ and $|s\rangle$ a set of fictitious spin-½ operators I_α^{rs} (α = x,y,z) is defined in the following way:

$$\begin{aligned} \langle i|I_x^{rs}|j\rangle &= \tfrac{1}{2}(\delta_{ir}\delta_{js} + \delta_{is}\delta_{jr}) \\ \langle i|I_y^{rs}|j\rangle &= \tfrac{1}{2}i(-\delta_{ir}\delta_{js} + \delta_{is}\delta_{jr}) \\ \langle i|I_z^{rs}|j\rangle &= \tfrac{1}{2}(\delta_{ir}\delta_{jr} - \delta_{is}\delta_{js}). \end{aligned} \qquad [III.7]$$

From this definition it follows that

$$\begin{aligned} I_x^{sr} &= I_x^{rs} \\ I_y^{sr} &= -I_y^{rs} \\ I_z^{sr} &= -I_z^{rs}. \end{aligned} \qquad [III.8]$$

The operators obey the conventional commutation rules:

$$[I_\alpha^{rs}, I_\beta^{rs}] = iI_\gamma^{rs}, \qquad [III.9]$$

where (α,β,γ) is a cyclic permutation of (x,y,z).
For two operators, defined for two connected transitions rt and st, one finds

$$\begin{aligned} [I_x^{rt}, I_x^{st}] &= [I_y^{rt}, I_y^{st}] = \tfrac{1}{2}iI_y^{rs} \\ [I_z^{rt}, I_z^{st}] &= 0 \\ [I_x^{rt}, I_y^{st}] &= \tfrac{1}{2}iI_x^{rs} \\ [I_x^{rt}, I_z^{st}] &= -\tfrac{1}{2}iI_y^{rt} \\ [I_y^{rt}, I_z^{st}] &= \tfrac{1}{2}iI_x^{rt}. \end{aligned} \qquad [III.10]$$

Operators defined for non-connected transitions rs and tu always commute,

$$[I_\alpha^{rs}, I_\beta^{tu}] = 0 \qquad \left.\begin{matrix}\alpha\\ \beta\end{matrix}\right\} = x,y,z. \qquad [III.11]$$

The mathematical forms [III.7]-[III.11] can be used for describing the effect of a selective pulse. A selective pulse $\theta_x(rs)$ with flip angle θ applied along the x-axis on a transition rs is analogous to Eq. [1.23] described by a rotation operator $R_x^{rs}(\theta)$, given by

$$R_x^{rs}(\theta) = \exp(-i\theta I_x^{rs}). \qquad [III.12]$$

A coherence present between eigenstates $|s\rangle$ and $|t\rangle$ only, is described by a density matrix σ which contains non-zero elements for σ_{st} and σ_{ts}, and can always be writen as

$$\sigma = C_x I_x^{st} + C_y I_y^{st}, \qquad [III.13]$$

where C_x and C_y are complex constants.

The $\theta_x(rs)$ pulse now creates a new density matrix σ', given by

$$\sigma' = \exp(i\theta I_x^{rs})\sigma \exp(-i\theta I_x^{rs}) = \qquad [III.14]$$

$$= \sum_{\alpha=x,y} C_\alpha \exp(i\theta I_x^{rs}) I_\alpha^{st} \exp(-i\theta I_x^{rs}).$$

Since it follows from Eq. [III.9] that

$$\exp(i\theta I_x^{rs}) I_x^{st} \exp(-i\theta I_x^{rs}) = \cos(\theta/2) I_x^{st} - \sin(\theta/2) I_y^{rt} \qquad [III.15]$$

and

$$\exp(i\theta I_x^{rs}) I_y^{st} \exp(-i\theta I_x^{rs}) = \cos(\theta/2) I_y^{st} + \sin(\theta/2) I_x^{rt}$$

the new density matrix σ' (Eq. [III.14]) can be written as

$$\sigma' = \cos(\theta/2)\sigma + \sin(\theta/2)\sigma'' \qquad [III.16]$$

with σ'' being a density matrix describing the newly created coherence between states $|r\rangle$ and $|t\rangle$:

$$\sigma'' = -C_x I_y^{rt} + C_y I_x^{rt}. \qquad [III.17]$$

Comparing the new density matrix σ' (Eq. [III.16]) with the initial density matrix σ (Eq. [III.13]) shows that a factor $\cos(\theta/2)$ of the initial coherence is still present, while new coherence rt is created with a magnitude proportional to $\sin(\theta/2)$.

Permutation of the order of selective pulses

In a weakly coupled N spin-½ system the operator for a non-selective pulse with flip angle θ, applied along the x-axis is given by

$$\hat{R}_x(\theta) = \exp(-i\theta\hat{F}_x). \qquad [III.18]$$

Since $\hat{F}_x = \sum_{n=1}^{N} \hat{I}_{x_n}$ and $[\hat{I}_{x_n},\hat{I}_{x_m}] = 0$ for $n \neq m$,

$$R_x(\theta) = \prod_{n=1}^{N} \exp(-i\theta\hat{I}_{x_n}), \qquad [III.19]$$

where the order of the product is arbitrary. $\hat{I}_{x_n}$ is the spin angular momentum operator along the x-axis for spin n, and is the sum of $2^{(N-1)}$ fictitious spin-½ operators I_x^{rs}, all working on different non-connected single quantum transitions, each corresponding to one of the multiplet components of nucleus n. These operators are labelled $I_{x_{n_m}}$ where $m = 1..2^{(N-1)}$. Since all of these operators commute (Eq.[III.11]), Eq.[III.19] can be written as

$$\hat{R}_x(\theta) = \prod_{n=1}^{N} \prod_{m=1}^{2^{(N-1)}} \exp(-i\theta\hat{I}_{x_{n_m}}). \qquad [III.20]$$

Each of the terms of Eq.[III.20] denotes a selective pulse, affecting only a single transition. The order of the selective pulses on nucleus n, denoted by $\exp(-i\theta\hat{I}_{x_{n_m}})$ can be randomly permuted since the operators $\hat{I}_{x_{n_m}}$ and $\hat{I}_{x_{n_p}}$ commute.

REFERENCES

(1) A. Wokaun and R.R. Ernst, J. Chem. Phys. 67, 1752 (1977).

(2) A. Wokaun, Thesis, E.T.H. Zürich, 1978.